# Fast Start Advanced Calculus

# Synthesis Lectures on Mathematics and Statistics

Editor
**Steven G. Kranz,** *Washington University, St. Louis*

Fast Start Advanced Calculus
Daniel Ashlock
2019

Fast Start Integral Calculus
Daniel Ashlock
2019

Fast Start Differential Calculus
Daniel Ashlock
2019

Introduction to Statistics Using R
Mustapha Akinkunmi
2019

Inverse Obstacle Scattering with Non-Over-Determined Scattering Data
Alexander G. Ramm
2019

Analytical Techniques for Solving Nonlinear Partial Differential Equations
Daniel J. Arrigo
2019

Aspects of Differential Geometry IV
Esteban Calviño-Louzao, Eduardo García-Río, Peter Gilkey, JeongHyeong Park, and Ramón Vázquez-Lorenzo
2019

Symmetry Problems. Thne Navier–Stokes Problem.
Alexander G. Ramm
2019

An Introduction to Partial Differential Equations
Daniel J. Arrigo
2017

Numerical Integration of Space Fractional Partial Differential Equations: Vol 2 –
Applicatons from Classical Integer PDEs
Younes Salehi and William E. Schiesser
2017

Numerical Integration of Space Fractional Partial Differential Equations: Vol 1 –
Introduction to Algorithms and Computer Coding in R
Younes Salehi and William E. Schiesser
2017

Aspects of Differential Geometry III
Esteban Calviño-Louzao, Eduardo García-Río, Peter Gilkey, JeongHyeong Park, and Ramón
Vázquez-Lorenzo
2017

The Fundamentals of Analysis for Talented Freshmen
Peter M. Luthy, Guido L. Weiss, and Steven S. Xiao
2016

Aspects of Differential Geometry II
Peter Gilkey, JeongHyeong Park, Ramón Vázquez-Lorenzo
2015

Aspects of Differential Geometry I
Peter Gilkey, JeongHyeong Park, Ramón Vázquez-Lorenzo
2015

An Easy Path to Convex Analysis and Applications
Boris S. Mordukhovich and Nguyen Mau Nam
2013

Applications of Affine and Weyl Geometry
Eduardo García-Río, Peter Gilkey, Stana Nikčević, and Ramón Vázquez-Lorenzo
2013

Essentials of Applied Mathematics for Engineers and Scientists, Second Edition
Robert G. Watts
2012

Chaotic Maps: Dynamics, Fractals, and Rapid Fluctuations
Goong Chen and Yu Huang
2011

Matrices in Engineering Problems
Marvin J. Tobias
2011

The Integral: A Crux for Analysis
Steven G. Krantz
2011

Statistics is Easy! Second Edition
Dennis Shasha and Manda Wilson
2010

Lectures on Financial Mathematics: Discrete Asset Pricing
Greg Anderson and Alec N. Kercheval
2010

Jordan Canonical Form: Theory and Practice
Steven H. Weintraub
2009

The Geometry of Walker Manifolds
Miguel Brozos-Vázquez, Eduardo García-Río, Peter Gilkey, Stana Nikčević, and Ramón Vázquez-Lorenzo
2009

An Introduction to Multivariable Mathematics
Leon Simon
2008

Jordan Canonical Form: Application to Differential Equations
Steven H. Weintraub
2008

Statistics is Easy!
Dennis Shasha and Manda Wilson
2008

A Gyrovector Space Approach to Hyperbolic Geometry
Abraham Albert Ungar
2008

Fast Start Advanced Calculus

Daniel Ashlock

ISBN: 978-3-031-01294-5     paperback
ISBN: 978-3-031-02422-1     ebook
ISBN: 978-3-031-00268-7     hardcover

DOI 10.1007/978-3-031-02422-1

A Publication in the Springer series
*SYNTHESIS LECTURES ON MATHEMATICS AND STATISTICS*

Lecture #30
Series Editor: Steven G. Kranz, *Washington University, St. Louis*
Series ISSN
Print 1938-1743    Electronic 1938-1751

# Fast Start Advanced Calculus

Daniel Ashlock
University of Guelph

*SYNTHESIS LECTURES ON MATHEMATICS AND STATISTICS #30*

## ABSTRACT

This book continues the material in two early Fast Start calculus volumes to include multivariate calculus, sequences and series, and a variety of additional applications. These include partial derivatives and the optimization techniques that arise from them, including Lagrange multipliers. Volumes of rotation, arc length, and surface area are included in the additional applications of integration. Using multiple integrals, including computing volume and center of mass, is covered. The book concludes with an initial treatment of sequences, series, power series, and Taylor's series, including techniques of function approximation.

## KEYWORDS

partial derivatives, multivariate optimization, constrained optimization, volume, arc-length, surface integrals, multiple integrams, series, power series

# Contents

# Preface

This text covers multi-variable integral and differential calculus, presuming familiarity with the single variable techniques from the precursor texts *Fast Start Differential Calculus* and *Fast Start Differential Calculus*. The texts were developed for a course that arose from a perennial complaint by the physics department at the University of Guelph that the introductory calculus courses covered topics roughly a year after they were needed. In an attempt to address this concern, a multi-disciplinary team created a two-semester integrated calculus and physics course. This book covers the integral calculus topics from that course as well a material on the behavior of polynomial function. The philosophy of the course was that the calculus will be delivered before it is needed, often just in time, and that the physics will serve as a substantial collection of motivating examples that will anchor the student's understanding of the mathematics.

The course has run three times before this text was started, and it was used in draft form for the fourth offering of the course, and then for two additional years. There is a good deal of classroom experience and testing behind this text. There is also enough information to confirm our hypothesis that the course would help students. The combined drop and flunk rate for this course is consistently under 3%, where 20% is more typical for first-year university calculus. Co-instruction of calculus and physics works. It is important to note that we did not achieve these results by watering down the math. The topics covered, in two semesters, are about half again as many as are covered by a standard first-year calculus course. That's the big surprise: covering more topics faster increased the average grade and reduced the failure rate. Using physics as a knowledge anchor worked even better than we had hoped.

This text, and its two companion volumes, *Fast Start Differential Calculus* and *Fast Start Integral Calculus*, make a number of innovations that have caused mathematical colleagues to raise objections. In mathematics it is traditional, even dogmatic, that math be taught in an order in which no thing is presented until the concepts on which it rests are already in hand. This is correct, useful dogma for mathematics students. It also leads to teaching difficult proofs to students who are still hungover from beginning-of-semester parties. This text neither emphasizes nor neglects theory, but it does move theory away from the beginning of the course in acknowledgment of the fact that this material is philosophically difficult and intellectually challenging. The course also presents a broad integrated picture as soon as possible. The text also emphasizes cleverness and computational efficiency. Remember that "mathematics is the art of avoiding calculation."

It is important to state what was sacrificed to make this course and this text work the way they do. This is not a good text for math majors, unless they get the theoretical parts of calculus later in a real analysis course. The text is relatively informal, almost entirely example driven, and application motivated. The author is a math professor with a CalTech Ph.D. and three decades of experience teaching math at all levels from 7th grade (as a volunteer) to graduate education including having supervised a dozen successful doctoral students. The author's calculus credits include calculus for math and engineering, calculus for biology, calculus for business, and multivariate and vector calculus.

Daniel Ashlock
August 2019

# Acknowledgments

This text was written for a course developed by a team including my co-developers Joanne M. O'Meara of the Department of Physics and Lori Jones and Dan Thomas of the Department of Chemistry, University of Guelph. Andrew McEachern, Cameron McGuinness, Jeremy Gilbert, and Amanda Saunders have served as head TAs and instructors for the course over the last six years and had a substantial impact on the development of both the course and this text. Martin Williams, of the Department of Physics at Guelph, has been an able partner on the physics side delivering the course and helping get the integration of the calculus and physics correct. I also owe six years of students thanks for serving as the test bed for the material. Many thanks to all these people for making it possible to decide what went into the text and what didn't. I also owe a great debt to Wendy Ashlock and Cameron McGuinness at Ashlock and McGuinness Consulting for removing a large number of errors and making numerous suggestions to enhance the clarity of the text. This is the fourth edition that corrects several mistakes and adds a very modest number of topics.

Daniel Ashlock
August 2019

# CHAPTER 1

# Advanced Derivatives

This chapter deals with the issues of derivatives of functions that have multiple independent variables. In our earlier studies we learned to take derivatives of functions of the form $y = f(x)$. Now we will learn to work with functions of the form $z = f(x, y)$, functions that graph as surfaces in a space like the one shown in Figure 1.1.

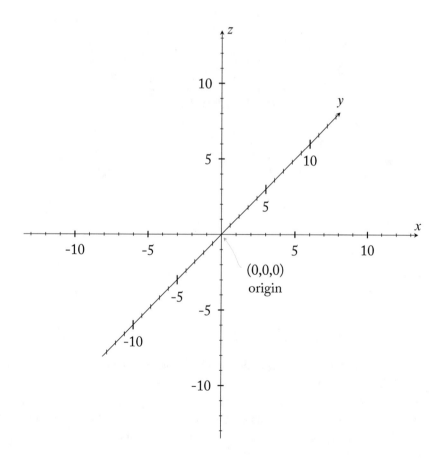

Figure 1.1: Three-dimensional coordinate system.

Where before we had points $(x, y)$, we now have points $(x, y, z)$. We've looked at the graph of functions like $y = x^2$ over and over. Now let's look at the three-dimensional analog:

$$z = x^2 + y^2$$

for $-4 \leq x, y \leq 4$. A graph of the function is shown in Figure 1.2.

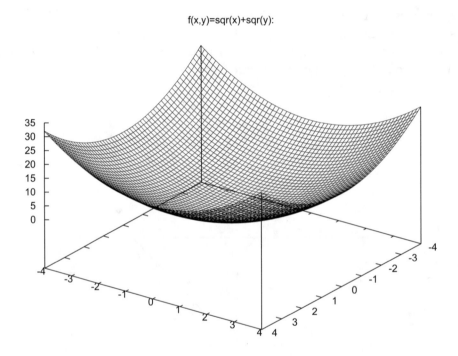

Figure 1.2: A graph of $f(x, y) = x^2 + y^2$ for $-4 \leq x, y \leq 4$.

## 1.1    PARTIAL DERIVATIVES

So far in this text we have had one independent and one dependent variable. Now we have two independent variables, which has a huge effect on derivatives. The number of directions on the number line is two – left and right or plus and minus. As we learned in *Fast Start Integral Calculus*, there are an infinite number of directions when there are two variables to choose directions among. Each unit vector starting at the origin points in a different direction. Every single vector can be written as:

$$\vec{v} = (a, b) = a \cdot (1, 0) + b \cdot (0, 1)$$

In other words, each vector is a combination of the two fundamental vectors $\vec{e}_1 = (1, 0)$ and $\vec{e}_2 = (0, 1)$.

On the surface of a function $z = f(x, y)$, the function has a rate of change in every direction at every point. Pick a direction – the slope in that direction is the rate of change in that direction. The basis for all of these rates of change are the *partial derivatives* – derivatives in the direction of $x$ and $y$.

**Knowledge Box 1.1**

**Partial Derivatives**

*If $z = f(x, y)$ is a function of two variables, then there are two fundamental derivatives*

$$z_x = \frac{\partial f}{\partial x} \text{ and } z_y = \frac{\partial f}{\partial y}$$

*These are called the* **partial derivatives** *of z (or f) with respect to x and y, respectively.*

In order to take the partial derivative of a function with respect to one variable, all other variables are treated as constants. This is most easily understood through examples.

**Example 1.1**    Suppose that $z = x^2 + 3xy + y^2$. Find $z_x$ and $z_y$.

**Solution:**

$$z_x = 2x + 3y \text{ and } z_y = 3x + 2y$$

To see this, notice that, when we are taking the derivative with respect to $x$, the derivative of $3xy$ is $3y$, *because $y$ is treated as a constant.* Similarly, the derivative of $3xy$ with respect to $y$ is $3x$. The derivatives for $x^2$ and $y^2$ are $2x$ and $2y$ when they are the active variable and zero when the other variable is active, because the derivative of a constant is zero.

$\Diamond$

It is important to remember that every derivative rule we have learned so far applies when we are taking partial derivatives – the product, quotient, and chain rules and all the individual formulas for functions.

**Example 1.2**   Find $\dfrac{\partial f}{\partial x}$ if

$$f(x, y) = \frac{x}{x^2 + y^2}$$

**Solution:**

This problem requires the quotient rule. Since a "prime" hash-mark doesn't carry its identity (with respect to $x$ or with respect to $y$) we used the symbol $\dfrac{\partial}{\partial x}$ to mean "derivative with respect to $x$."

This means that

$$\frac{\partial}{\partial x}\left(\frac{x}{x^2 + y^2}\right) = \frac{\frac{\partial}{\partial x}x \cdot (x^2 + y^2) - \frac{\partial}{\partial x}(x^2 + y^2) \cdot x}{(x^2 + y^2)^2}$$

$$= \frac{1 \cdot (x^2 + y^2) - 2x \cdot x}{(x^2 + y^2)^2}$$

$$= \frac{y^2 - x^2}{(x^2 + y^2)^2}$$

$$\Diamond$$

The next example uses the chain rule. It also uses the additional notation "$f_y$" as another way of saying "the partial derivative of $f(x, y)$ with respect to $y$."

**Example 1.3**   Find $f_y(x, y)$ if

$$f(x, y) = \sin(2xy + 1)$$

**Solution:**

$$f_y(x, y) = \cos(2xy + 1) \cdot 2x$$

$$\Diamond$$

As long as we remember that $y$ is the active variable and $x$ is treated as a constant, this is not difficult.

**Example 1.4**   Find the partial derivatives with respect to $x$ and $y$ of

$$f(x, y) = (3x + 4y)^5$$

**Solution:**

$$f_x(x, y) = 5(3x + 4y)^4 \cdot 3 \qquad f_y(x, y) = 5(3x + 4y)^4 \cdot 4$$

The only point where things are different is the way the chain rule acts depending on if $x$ or $y$ is the active variable.

$\diamond$

**Example 1.5**   Find the partial derivatives with respect to $x$ and $y$ for

$$f(x, y) = \frac{x}{y}$$

**Solution:**

$$f_x(x, y) = \frac{1}{y} \qquad f_y(x, y) = \frac{-x}{y^2}$$

For $f_x$, $\dfrac{1}{y}$ is effectively a constant, while, for $f_y$, we employ the reciprocal rule while $x$ plays the part of a constant.

$\diamond$

### 1.1.1   IMPLICIT PARTIAL DERIVATIVES

Since natural laws are often stated in the form of equations that are not in functional form, implicit derivatives are very useful in physics. It turns out that implicit partial derivatives are a lot like standard partial derivatives – as long as you remember which variable is active.

**Example 1.6**   Find $z_x$ and $z_y$ if $x^2 + y^2 + z^2 = 16$.

**Solution:**

$z$ is the dependent variable and so gets a $z_x$ or a $z_y$ each time we take a derivative of it, while $x$ and $y$ take turns being the active variable and a constant, respectively. So,

$$2x + 2z \cdot z_x = 0 \text{ and } 2y + 2z \cdot z_y = 0$$

Simplifying we get that

$$z_x = -\frac{x}{z} \text{ and } z_y = -\frac{y}{z}$$

◊

**Example 1.7**   Find $z_y$ if

$$(xyz + 1)^3 = 6y$$

**Solution:**

In this example $z$ is the dependent variable, $y$ is the active variable, and $x$ is acting like a constant. So:

$$3(xyz + 1)^2(xz + xy \cdot z_y) = 6$$

$$xz + xy \cdot z_y = \frac{2}{(xyz + 1)^2}$$

$$z_y = \frac{2}{xy \cdot (xyz + 1)^2} - \frac{z}{y}$$

◊

## 1.1.2   HIGHER-ORDER PARTIAL DERIVATIVES

In earlier chapters, when we wanted the second derivative of a function we just took the derivative again. The fact that we have multiple choices of which derivative to take complicates this. If the first derivative is with respect to $x$ or $y$ and so is the second, then we get four possible nominal second derivatives:

$$f_{xx} \qquad f_{xy} \qquad f_{yx} \qquad f_{yy}$$

A useful fact saves us from having the number of higher-order partial derivatives explode.

**Knowledge Box 1.2**

*When working with $f(x, y)$,*

$$f_{xy}(x, y) = f_{yx}(x, y)$$

*and in general the order in which partial derivatives are taken does not affect the result of taking them.*

**Example 1.8** If

$$f(x, y) = x^3 + 3xy + y^2$$

find $f_{xx}$, $f_{xy}$, $f_{yx}$, and $f_{yy}$, and verify that $f_{xy} = f_{yx}$.

**Solution:**

Start by finding $f_x$ and $f_y$ and then keep going.

$$f_x = 3x^2 + 3y$$

$$f_y = 3x + 2y$$

$$f_{xx} = 6x$$

$$f_{xy} = 3$$

$$f_{yx} = 3$$

$$f_{yy} = 2$$

And we see that $f_{xy} = 3 = f_{yx}$, achieving the desired verification. From this point on we will only compute one of the two mixed partials $f_{xy}$ and $f_{yx}$.

◊

The "order doesn't matter" rule means, for example, that $f_{xxy} = f_{xyx} = f_{yxx}$. So the degree to which this rule reduces the number of higher-order derivatives increases with the order of the derivative.

**Example 1.9** Find $f_{xx}$, $f_{xy}$, and $f_{yy}$ for

$$f(x, y) = x^2 \sin(y).$$

**Solution:**

Compute the first partials first and keep going.

$$f_x = 2x \sin(y)$$

$$f_y = x^2 \cos(y)$$

$$f_{xx} = 2 \sin(y)$$

$$f_{xy} = 2x \cos(y)$$

$$f_{yy} = -x^2 \sin(y)$$

$$\Diamond$$

**Example 1.10**   Find $f_{xy}$ for

$$f(x, y) = \frac{xy}{x^2 + 1}.$$

**Solution:**

The fact that we get the same result by computing the partials with respect to $x$ and $y$ in either order means that we may choose the order to minimize our work. The order $y$ then $x$ is easier, because $y$ vanishes.

$$f(x, y) = y \cdot \frac{x}{x^2 + 1}$$

$$f_y = \frac{x}{x^2 + 1}$$

$$f_{xy} = \frac{(x^2 + 1)(1) - x(2x)}{(x^2 + 1)^2} \qquad\qquad \text{Quotient rule.}$$

$$= \frac{1 - x^2}{(x^2 + 1)^2}$$

$$\Diamond$$

# PROBLEMS

**Problem 1.11**    For each of the following functions find $f_x$ and $f_y$.

1. $f(x, y) = 2x^2 + 3xy + y^2 + 4x + 2y + 7$

2. $g(x, y) = \sin(xy)$

3. $h(x, y) = x^3 y^3$

4. $r(x, y) = \ln(x^2 + y^2 + 1)$

5. $s(x, y) = \dfrac{1}{x^2 + y^2 + 1}$

6. $q(x, y) = \dfrac{x - y}{x + y}$

7. $a(x, y) = e^{x \sin(y)}$

8. $b(x, y) = (x^2 + xy + 1)^6$

**Problem 1.12**    For each of the following functions find $f_{xx}$, $f_{xy}$, and $f_{yy}$.

1. $f(x, y) = x^2 - 5xy + 2y^2 + 3x - 6y + 11$

2. $g(x, y) = \tan^{-1}(xy)$

3. $h(x, y) = (x^3 + 1)(y^3 + 1)$

4. $r(x, y) = e^{x^2 + y^2 - 5}$

5. $s(x, y) = \dfrac{1}{x^2 + y^2 + 1}$

6. $q(x, y) = \dfrac{2x - 3y}{5x + y}$

7. $a(x, y) = \ln(x \sin(y))$

8. $b(x, y) = (2x^2 - xy)^4$

**Problem 1.13**    Find $z_x$ and $z_y$ if $(xyz + 2)^3 = 4$.

**Problem 1.14**    Find $z_x$ and $z_y$ if $\dfrac{x}{yz} = 1$.

**Problem 1.15**    Find $z_x$ and $z_y$ if $\dfrac{x + z}{y + z} = 4$.

**Problem 1.16**    Find $z_x$ and $z_y$ if $\dfrac{3x + z}{2z} = 4xy$.

**Problem 1.17**    Find $z_x$ and $z_y$ if $\cos(x + y + z) = \dfrac{\sqrt{2}}{2}$.

**Problem 1.18**    Find $z_x$ and $z_y$ if $\tan^{-1}(z - xy) = 1$.

**Problem 1.19**    Find $z_x$ and $z_y$ if $(xy + xz + yz)^3 = 16$.

**Problem 1.20**   For each of the following functions find $f_{xy}$.

1. $f(x, y) = x^2y + x^3y + y\sin(x)$

2. $g(x, y) = x\sin(y) + x\tan^{-1}(x)$

3. $h(x, y) = xy^2 + x^2y + xy$

4. $r(x, y) = \left(x(y + 1)^5 + 1\right)^2$

5. $s(x, y) = x \cdot \sin(xy)$

6. $q(x, y) = y \cdot \cos(xy)$

7. $a(x, y) = \dfrac{x}{y} + \dfrac{y}{x} + y^3$

8. $b(x, y) = (1 + x + y + xy)^4$

**Problem 1.21**   Find $f_x$, $f_y$, $f_{xx}$, $f_{xy}$, and $f_{yy}$ if $f(x, y) = \dfrac{x^2}{x^2 + y^2}$.

**Problem 1.22**   Find $g_x$, $g_y$, $g_{xx}$, $g_{xy}$, and $g_{yy}$ if $g(x, y) = \tan^{-1}(xy + 1)$.

**Problem 1.23**   Rind $f_{xx}$, $f_{xxy}$, and $f_{xxyy}$ if $h(x, y) = \sin(xy)$.

**Problem 1.24**   Find $z_x$ and $z_y$ if $z = (x^2 + 1)^y$.

**Problem 1.25**   Find $z_x$ and $z_y$ if $z^{xy} = 2$.

**Problem 1.26**   Find $z_x$ and $z_y$ if $z = \dfrac{x^3(x + 1)^2(x - 1)^3}{y^2(y - 1)^3(y + 1)^5}$.

## 1.2   THE GRADIENT AND DIRECTIONAL DERIVATIVES

We are now ready to look at some of the opportunities that are available once we understand partial derivatives. At any point on the surface that forms the graph of $z = f(x, y)$ there are an infinite number of directions and so an infinite number of rates at which the function is changing. Pick a direction, and the function has a rate of change *in that direction*.

It turns out that there is some order to this richness of directions and rates of growth, in the form of a simple formula for the direction in which the function is growing fastest.

**Knowledge Box 1.3**

*If $z = f(x, y)$ is a function of two variables, then*

$$\nabla f(x, y) = \left( f_x, f_y \right)$$

*is called the* **gradient** *of $f(x, y)$. The gradient of a function points in the direction it is growing most quickly; the rate of growth is the magnitude of the gradient.*

**Example 1.27**    Find the gradient of the function $f(x, y) = x^2 + y^2 + 3xy$.

**Solution:**

Using the formula given, $\nabla f(x, y) = (2x + 3y, 3x + 2y)$.

◇

We can ask much more complex questions about the gradient than simply computing its value.

**Example 1.28**    At what points is the function

$$g(x, y) = \sin(x) + \cos(y)$$

changing the fastest in its direction of maximum increase?

**Solution:**

This question wants us to maximize the magnitude of the gradient.

First compute the gradient:

$$\nabla g(x, y) = (\cos(x), -\sin(y))$$

The magnitude of this is

$$\sqrt{\cos^2(x) + \sin^2(y)}$$

Since $x$ and $y$ vary independently, the answer is simply those points that make $\cos^2(x)$ and $\sin^2(y)$ both one.

So, the answer is those points $(x, y)$ such that

$$x = n\pi \text{ and } y = \frac{2m + 1}{2}\pi$$

where $n$ and $m$ are whole numbers.

◇

The graph in Figure 1.3 might help you understand Example 1.28.

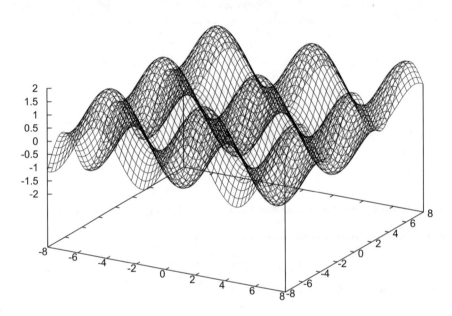

Figure 1.3: A graph of $z = \sin(x) + \cos(y)$ for $-8 \leq x, y \leq 8$.

**Example 1.29**  Find the gradient of

$$z = x^y$$

**Solution:**

The partial derivative with respect to $x$ is not hard because $y$ is treated as a constant – so $z_x = yx^{y-1}$. The partial derivative with respect to $y$ is trickier. It uses the formula for the derivative of a constant to a variable power which gives $z_y = x^y \cdot \ln(x)$.

This makes the gradient

$$\nabla z = \left(yx^{y-1}, x^y \cdot \ln(x)\right)$$

◇

What is the physical meaning of the gradient? We have already noted that it points in the direction in which a surface grows fastest away from the point where the gradient is computed – the steepest uphill slope away from the point. The magnitude of the gradient also gives of the *rate* of fastest growth. Another fact is that the negative of the function

$$-\nabla f(x, y)$$

is the steepest downhill slope away from the point.

In other words, the negative of the gradient is the direction that a ball, starting at rest, will roll.

f(x,y)=2xy/(x^2+y^2+1)

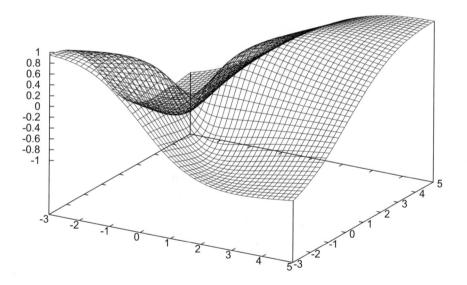

Figure 1.4: A graph of $H(x, y) = \dfrac{2x + y}{x^2 + y^2 + 1}$ for $-3 \le x, y \le 5$

**Example 1.30**   Suppose that the function $H(x, y) = \dfrac{2x + y}{x^2 + y^2 + 1}$ (Figure 1.4) describes the height of a surface. Which direction will a ball placed at the point $(1, 2)$ roll?

**Solution:**

We need to compute the negative of the gradient of $H(x, y)$.

$$H_x(x, y) = \frac{(x^2 + y^2 + 1)(2) - (2x + y)(2x)}{(x^2 + y^2 + 1)^2}$$

$$H_x(1, 2) = \frac{12 - 8}{36} = 1/9$$

$$H_y(x, y) = \frac{(x^2 + y^2 + 1)(1) - (2x + y)(2y)}{(x^2 + y^2 + 1)^2}$$

$$H_y(1, 2) = \frac{6 - 8}{36} = -1/18$$

So the ball rolls in the direction of the vector

$$-\nabla H(1, 2) = \left(\frac{-1}{9}, \frac{1}{18}\right)$$

◇

Notice that, if we were designing a game, then we could use a well-chosen equation to give us a height map for the surface, and the gradient could be used to tell which ways balls would roll and water would flow.

Visualization helps us understand functions. This leads to the question: what does a gradient look like? The gradient of a function $f(x, y)$ assigns a vector to each point in space. That means that we could get an idea of what a gradient looks like by plotting the vectors of the gradient on a grid of points.

**Example 1.31**   Let $f(x, y) = x^2 + y^2$. For all points $(x, y)$ with coordinates in the set $\{\pm 2, \pm 1.5, \pm 1, \pm 0.5, 0\}$, plot the point and the vector starting at that point in the direction $\nabla f(x, y)$.

**Solution:**

Since $\nabla f(x, y) = (2x, 2y)$ the vectors are:

2.5

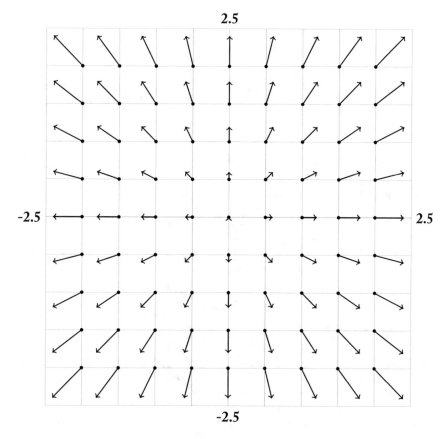

-2.5                                                                    2.5

-2.5

Notice that all the gradient vectors point directly away from the origin. This corresponds to our notion that the gradient is the direction of fastest growth.

◊

The function $\dfrac{2x + y}{x^2 + y^2 + 1}$ from Example 1.30, the rolling ball question, will probably have a more interesting gradient than the simple paraboloid in Example 1.31.

**Example 1.32**   Let $H(x, y) = \dfrac{2x + y}{x^2 + y^2 + 1}$. For all points $(x, y)$ with coordinates in the set $\{\pm 3, \pm 2.5, \pm 2, \pm 1.5, \pm 1, \pm 0.5, 0\}$, plot the point and the vector starting at that point in the direction $\nabla H(x, y)$.

**Solution:**

We are plotting the vectors drawn from the gradient

$$\nabla H(x, y) = \left( \frac{2y^2 - 2x^2 - 2xy + 2}{(x^2 + y^2 + 1)^2}, \frac{x^2 - y^2 - 4xy + 1}{(x^2 + y^2 + 1)^2} \right)$$

which yields the picture:

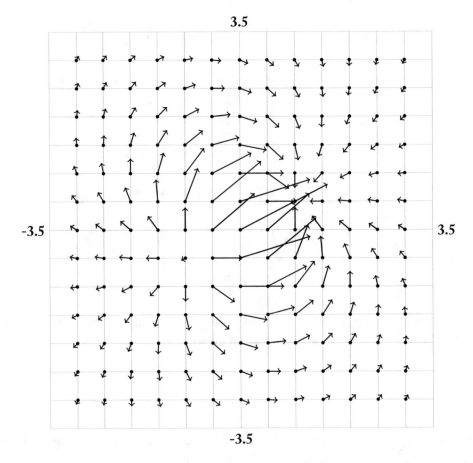

The gradient vectors point in many directions. A ball rolling on this surface could potentially have a very complex path.

◇

**Example 1.33**    Find a reasonable sketch of the vector field associated with the gradient of $g(x, y) = \sin(x) + \cos(y)$ from Example 1.28. Use grid points with coordinates $\pm n$ for $n = 0, 1, \ldots 8$.

**Solution:**
We are plotting the vectors drawn from the gradient

$$\nabla g(x, y) = (\cos(x), -\sin(y))$$

which yields the picture:

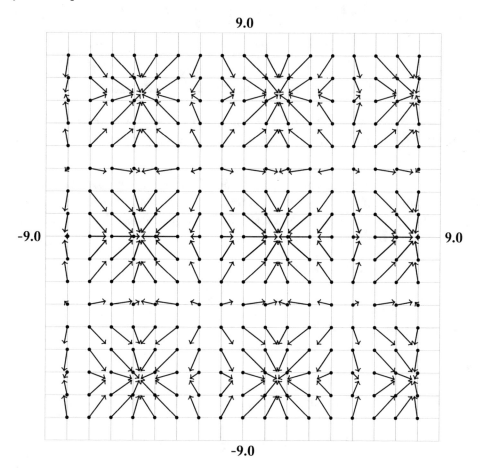

The periodicity of the vector field is easy to see. The different cells of the periodicity are slightly different because the periods are multiples of $\pi$, and the grid used to display the vectors is sized in multiples of one.

◊

Now that we have the gradient, it is possible to define the derivative in a particular direction.

### Knowledge Box 1.4

*If $z = f(x, y)$ is a differentiable function of two variables, and $\vec{u} = (a, b)$ is a unit vector, then the* **derivative of $f(x, y)$ in the direction of $\vec{u}$ is:**

$$\nabla_{\vec{u}} f(x, y) = \vec{u} \cdot \nabla f(x, y) = a f_x + b f_y$$

*If $\vec{v} = (r, s)$ is any vector, then the* **derivative of $f(x, y)$ in the direction of $\vec{v}$ is:**

$$\nabla_{\vec{v}} f(x, y) = \frac{\vec{v}}{|\vec{v}|} \cdot \nabla f(x, y)$$

Notice that we are continuing the practice of using unit vectors to designate directions – even when computing the derivative of a function in the direction of a general vector, we first coerce it to be a unit vector.

It is worth mentioning that, when computing directional derivatives, we start with a scalar quantity – the function $f(x, y)$. When we compute the gradient, we get the vector quantity $\nabla f(x, y) = (f_x, f_y)$ but then return to a scalar function of two variables $\nabla_{\vec{u}} f(x, y)$. It is important to keep track of the type of object – scalar or vector – that you are working with.

**Example 1.34**   Find the derivative of

$$f(x, y) = x^2 + y^2$$

in the direction of $\vec{u} = (1/2, \sqrt{3}/2)$.

**Solution:**

The vector $\vec{u}$ is a unit vector so, starting with $f_x = 2x$, $f_y = 2y$, we get:

$$\nabla_{\vec{u}} f(x, y) = \frac{1}{2} 2x + \frac{\sqrt{3}}{2} 2y = x + \sqrt{3} y$$

$$\Diamond$$

The directional derivative occasionally comes up in the natural course of trying to solve a problem. There is one very natural application: finding level curves. First let's define level curves.

**Knowledge Box 1.5**

**Level Curves**

*If $z = f(x, y)$ defines a surface, then the **level curve of height c** of $f(x, y)$ is the set of points that solve the equation*

$$f(x, y) = c.$$

**Example 1.35**   Plot the level curves for $c \in \{1, 2, 3, 4, 5, 6\}$ for

$$f(x, y) = x^2 + 2y^2$$

**Solution:**

The equation $x^2 + 2y^2 = c$ is an ellipse that is $\sqrt{2}$ times as far across in the $x$ direction as the $y$ directions. Solving for the points where $x = 0$ or $y = 0$ gives us the extreme points of the ellipse for each value of $c$, and we get the following picture.

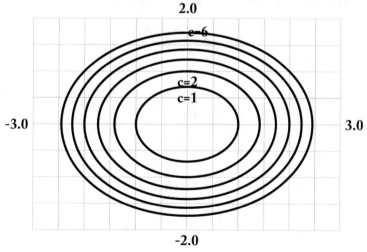

Level curves for $f(x, y) = x^2 + 2y^2$.

Since the values of $c$ used to derive the level curves are equally spaced, the fact that the curves are getting closer together gives a sense of how steeply the graph of $f(x, y) = x^2 + 2y^2$ is sloped.

◇

In Example 1.35 we can plot the level curves because we recognize them as ellipses. In general, that's not going to work. If we supply a graphics system with information about the gradient of a curve, then it can sketch level curves by following the correct directional derivative.

### Knowledge Box 1.6

*If $z = f(x, y)$ defines a surface, then the level curve at the point $(a, b)$ proceeds in the direction of the unit vector $\vec{u}$ such that*

$$|\nabla_{\vec{u}} f(a, b)| = 0.$$

*In other words, it proceeds in the direction such that the height of the graph of $f(x, y)$ is not changing.*

**Example 1.36**   Find, in general, the direction of the level curves of

$$f(x, y) = x^2 + 3y^2.$$

**Solution:**

To find the general direction of level curves, we need to solve this equation for $\vec{u}$.

$$|\nabla_{\vec{u}} f(a, b)| = 0$$

Let $\vec{u} = (a, b)$.

$$\nabla \left( x^2 + 3y^2 \right) \cdot (a, b) = 0$$

$$(2x, 6y) \cdot (a, b) = 0$$

$$2xa + 6yb = 0$$

$$2xa = -6yb$$

$$a = -3 \left( \frac{y}{x} \right) b$$

Giving us the direction in which the level curves go.

If we set $b = 1$, then, at a given point $(x, y)$ in space, the level curve at that point is in the direction $\left(-\dfrac{3y}{x}, 1\right)$. This encodes the direction.

To be thorough, let's turn this into a unit vector.

$$\left|\left(-\frac{3y}{x}, 1\right)\right| = \sqrt{\left(-\frac{3y}{x}\right)^2 + 1^2} = \sqrt{\frac{9y^2}{x^2} + 1} = \frac{1}{x}\sqrt{9y^2 + x^2}$$

So, the direction, as a unit vector is:

$$\left(-\frac{3y}{\sqrt{9y^2 + x^2}}, \frac{x}{\sqrt{9y^2 + x^2}}\right)$$

Notice that if we had chosen $b = -1$, we would have gotten

$$\left(\frac{3y}{\sqrt{9y^2 + x^2}}, -\frac{x}{\sqrt{9y^2 + x^2}}\right)$$

which is an equally valid solution.

The level curve points in two opposite directions.

◊

A natural question at this point is: what do the vectors we found in Example 1.36 look like?

**Example 1.37**  Using the solution to Example 1.36 plot the directions of the level curves as a vector field.

**Solution:**

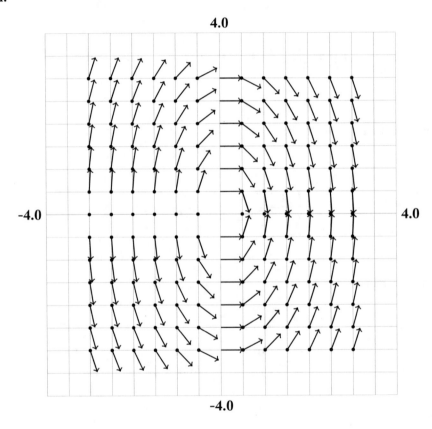

At each point, there are actually two vectors that point in the direction of the level curve. If $\vec{u}$ points in the direction of the level curve, then so does $-\vec{u}$. The figure above sometimes shows one and sometimes the other. The curves that these vectors are following track the vertically long ellipses that one would expect as level curves of:

$$f(x, y) = x^2 + 3y^2$$

◊

# PROBLEMS

**Problem 1.38**    For each of the following functions, find their gradient.

1. $f(x, y) = x^2 + 4xy + 7y^2 + 2x - 5y + 1$

2. $g(x, y) = \sin(xy)$

3. $h(x, y) = \dfrac{x^2}{y^2 + 1}$

4. $r(x, y) = (x^2 + y^2)^{3/2}$

5. $s(x, y) = \sin(x) + \cos(y)$

6. $q(x, y) = x^5 y + xy^5 + x^3 y^3 + 1$

7. $a(x, y) = \dfrac{x}{x^2 + y^2}$

8. $b(x, y) = x^y$

**Problem 1.39**    For the function

$$g(x, y) = x^2 + 3xy + y^2,$$

remembering that directions should be reported as unit vectors, find:

1. The greatest rate of growth of the curve in any direction at the point $(1, 1)$

2. The direction of greatest growth at $(1, 1)$

3. The greatest rate of growth of the curve in any direction at the point $(-1, 2)$

4. The direction of greatest growth at $(-1, 2)$

5. The greatest rate of growth of the curve in any direction at the point $(0, 3)$

6. The direction of greatest growth at $(0, 3)$

**Problem 1.40**    Suppose we are trying to find level curves. Is it possible to find points where there are more than two directions in which the surface does not grow? Either explain why this cannot happen or give an example where it does.

**Problem 1.41**    For each of the following functions, sketch the gradient vector field of the function for the points $(x, y)$ with

$$x, y \in \{0, \pm 1, \pm 2, \pm 3\}.$$

Exclude points, if any, where the gradient does not exist. This is a problem where a spreadsheet may be useful for performing routine computation.

1. $f(x, y) = x^2 + y^2 + 1$

2. $g(x, y) = \sin(xy)$

3. $h(x, y) = \dfrac{x}{x + y}$

4. $r(x, y) = \sqrt{x^2 + y^2}$

5. $s(x, y) = \sin(x) + \cos(y)$

6. $q(x, y) = x^5 y + xy^5 + x^3 y^3 + 1$

**Problem 1.42**   Find the directional derivative of

$$f(x, y) = x^2 + y^3$$

in the direction of (1,1) at (0,0).

**Problem 1.43**   Find the directional derivative of

$$g(x, y) = x^2 - y^2$$

in the direction of $(\sqrt{3}/2, 1/2)$ at (2,1).

**Problem 1.44**   Find the directional derivative of

$$h(x, y) = (y - 2x)^3$$

in the direction of (0,-1) at (-1,1).

**Problem 1.45**   As in Example 1.36 find the general direction for level curves at $(x, y)$ for the function

$$H(x, y) = \frac{1}{x^2 + y^2}$$

and plot the vector field for all points with whole number coordinates in the range $-3 \le x, y \le 3$.

**Problem 1.46**   As in Example 1.36 find the general direction for level curves at $(x, y)$ for the function

$$H(x, y) = \frac{2}{x^2 - 2x + y^2 - 2y + 2}$$

and plot the vector field for all points with whole number coordinates in the range $-5 \le x, y \le 5$.

**Problem 1.47**   Which way will a ball placed at (1,1) on the surface given by the function in Problem 1.42 roll?

## 1.3    TANGENT PLANES

One of the first things we built after developing skill with the derivative was tangent lines to a curve. With a function $z = f(x, y)$ the analogous object is a tangent *plane*. There are several ways to specify a plane.

**Knowledge Box 1.7**

**Formulas for planes**

*If a, b, c, and d are constants, all of the following formulas specify planes:*

$$z = ax + by + c$$

$$ax + by + cz = d.$$

$$(a, b, c) \cdot (x, y, z) = d$$

*Note that the third formula, while phrased in terms of a dot product, is actually the same as the second.*

Let's get some practice with converting between the different possible forms of a plane.

**Example 1.48**    If

$$(3, -2, 5) \cdot (x, y, z) = 2$$

find the other two forms of the plane.

**Solution:**

$$
\begin{aligned}
(3, -2, 5) \cdot (x, y, z) &= 2 \\
3x - 2y + 5z &= 2 \qquad\qquad \text{Second form} \\
5z &= -3x + 2y + 2 \\
z &= -0.6x + 0.4y + 0.4 \qquad\qquad \text{First form}
\end{aligned}
$$

$\Diamond$

The most common way to find a tangent line for a function is to take the point of tangency – which must be on the line – together with the slope of the line found by computing the derivative and use the point slope form to find the formula for the line. It turns out that there is a way of specifying planes that is similar to the point slope form of a line. Remember that if $\vec{v}$ and $\vec{w}$ are vectors that are at right angles to one another, then $\vec{v} \cdot \vec{w} = 0$.

**Knowledge Box 1.8**

### Formula for a plane at right angles to a vector

*Suppose that $\vec{v}$ is a vector at right angles to a plane in three dimensions and that $(a, b, c)$ is a point on that plane. Then a formula for the plane is*

$$\vec{v} \cdot (x - a, y - b, z - c) = 0.$$

*Notice that $(a, b, c)$ is constructively on the plane and that our knowledge of the dot product tells us it is at right angles to $\vec{v}$.*

**Example 1.49**   Find the plane at right angles to $\vec{v} = (1, 2, 1)$ through the point $(2, -1, 5)$.

**Solution:**

Simply substitute into the formula given in Knowledge Box 1.8.

$$
\begin{aligned}
(1, 2, 1) \cdot (x - 2, y + 1, z - 5) &= 0 \\
1(x - 2) + 2(y + 1) + 1(z - 5) &= 0 \\
x + 2y + z - 2 + 2 - 5 &= 0 \\
x + 2y + z &= 5
\end{aligned}
$$

$\Diamond$

Knowledge Box 1.8 seems very special purpose, but it turns out to be very useful in light of another fact. Suppose that

$$f(x, y, z) = c$$

specifies a surface. We need to expand our definition of the gradient just a bit to

$$\nabla f(x, y, z) = \left( f_x(x, y, z), f_y(x, y, z), f_z(x, y, z) \right).$$

In this case the vector $\nabla f(a, b, c)$ points directly outward from the surface $f(x, y, z) = c$ at $(a, b, c)$ and *it is at right angles to the tangent plane.*

**Knowledge Box 1.9**

### Formula for the tangent plane to a surface

*If $f(x, y, z) = c$ defines a surface, and $(a, b, c)$ is a point on the surface, then a formula for the tangent plane to that surface at $(a, b, c)$ is:*

$$\left( f_x(a, b, c), f_y(a, b, c), f_z(a, b, c) \right) \cdot (x - a, y - b, z - c) = 0.$$

Defining a surface in the form $f(x, y, z) = c$ is a little bit new – but in fact this is another version of level curves, just one dimension higher. Let's practice.

**Example 1.50**    Find the tangent plane to the surface $x^2 + y^2 + z^2 = 3$ at the point $(1, -1, 1)$.

**Solution:**

Check that the point is on the surface: $(1)^2 + (-1)^2 + (1)^2 = 3$ – so it is. Next find the gradient

$$\nabla x^2 + y^2 + z^2 = (2x, 2y, 2z).$$

This means that the gradient at $(1, -1, 1)$ is $\vec{v} = (2, -2, 2)$. This makes the plane

$$
\begin{aligned}
(2, -2, 2) \cdot (x - 1, y + 1, z - 1) &= 0 \\
2x - 2y + 2z - 2 - 2 - 2 &= 0 \\
2x - 2y + 2z &= 6 \\
x - y + z &= 3
\end{aligned}
$$

Notice that we simplified the form of the plane; this is not required but it does make for neater answers.

$\Diamond$

The problem with finding the tangent plane to a surface $f(x, y, z) = c$ is that it does not solve the original problem – finding tangent planes to $z = f(x, y)$. A modest amount of algebra solves this problem. If $z = f(x, y)$ then $g(x, y, z) = z - f(x, y) = 0$ is in the correct form for our surface techniques. This gives us a new way of finding tangent planes to a function that defines a surface in 3-space.

**Knowledge Box 1.10**

**Formula for the tangent plane to a functional surface**

*If $z = f(x, y)$ defines a surface, then the tangent plane to the surface at $(a, b)$ may be obtained as*

$$\left(-f_x(a, b), -f_y(a, b), 1\right) \cdot (x - a, y - b, z - f(a, b)) = 0.$$

*This is the result of applying the gradient-of-a-surface formula to the surface $z - f(x, y) = 0$ at the point $(a, b, f(a, b))$.*

**Example 1.51**    Find the tangent plane to $f(x, y) = x^2 - y^3$ at the point $(2, -1)$.

**Solution:**

Assemble the pieces and plug into Knowledge Box 1.10.

$$f_x(x, y) = 2x$$

$$f_x(2, -1) = 4$$

$$f_y(x, y) = -3y^2$$

$$f_y(2, -1) = -3$$

$$f(2, -1) = 4 - (-1) = 5$$

Put the plane together

$$(-4, 3, 1) \cdot (x - 2, y + 1, z - 5) = 0$$

$$-4x + 8 + 3y + 3 + z - 5 = 0$$

$$-4x + 3y + z = -6$$

Which is the tangent plane desired.

◊

**Example 1.52**   Find the tangent plane to

$$g(x, y) = xy^2$$

at (3,1).

**Solution:**

Assemble the pieces and plug into Knowledge Box 1.10.

$$f_x(x, y) = y^2$$

$$f_x(3, 1) = 1$$

$$f_y(x, y) = 2xy$$

$$f_y(3, 1) = 6$$

$$f(3, 1) = 3$$

Put the plane together

$$(-1, -6, 1) \cdot (x - 3, y - 1, z - 3) = 0$$

$$-x + 3 - 6y + 6 + z - 3 = 0$$

$$-x - 6y + z = -6$$

$$\Diamond$$

# PROBLEMS

**Problem 1.53**   Find the plane through (1,-1,1) at right angles to $\vec{v} = (2, 2, -1)$.

**Problem 1.54**   Find the plane through (2,0,5) at right angles to $\vec{v} = (1, 1, 1)$.

**Problem 1.55**   Find the plane through (3,2,1) at right angles to $\vec{v} = (1, -1, 2)$.

**Problem 1.56**   Find, in the form

$$ax + by + cz = d,$$

the tangent planes to the following curves at the indicated points.

1. $f(x, y) = x^2 + y^2 - 1$ at $(2, 2)$

4. $r(x, y) = y \cdot \ln(x^2 + 1)$ at $(3, -1)$

2. $g(x, y) = 3x^2 + 2xy + 4^2 - 1$ at $(-1, 1)$

5. $s(x, y) = e^{x^2 + y^2}$ at $(0, 1)$

3. $h(x, y) = (x + y + 1)^2$ at $(0, 4)$

6. $q(x, y) = \sin(x) \cos(y)$ at $(\pi/3, \pi/6)$

**Problem 1.57**   Suppose that

$$x^2 + y^2 + z^2 = 12.$$

Find the tangent plane at each of the following points.

1. $p = (2, 2, 2)$

3. $r = (-2, -2, -2)$

5. $v = (-\sqrt{7}, 1, 2)$

2. $q = (2, -2, 2)$

4. $u = (1, 1, \sqrt{10})$

6. $w = (-1/2, 2.5, \sqrt{22}/2)$

**Problem 1.58**   Find all tangent planes to

$$x^2 + y^2 + z^2 = 75$$

that are at right angles to the vector $(2, 4, 6)$.

**Problem 1.59**   Find the tangent plane at $(-1, \pi/3)$ to

$$g(x, y) = x \cos(y)$$

**Problem 1.60**   Find the tangent plane to each of the following surfaces at the indicated point.

1. Surface $x + y + z^2 = 6$ at $(1,1,2)$

2. Surface $x^2 - y^3 + 5z = 19$ at $(4,-2,-1)$

3. Surface $3x + y^2 + z^2 = 8$ at $(2,1,1)$

4. Surface $(x - y)^3 + 2z = 14$ at $(2,0,3)$

5. Surface $xyz + x + y + z = 4$ at $(1,1,1)$

6. Surface $xy + yz + xz = 0$ at $(-1,1,-2)$

7. Surface $x^2 + y^2 + z^3 = 6$ at $(2,1,1)$

8. Surface $xy + xz + yz = 7$ at $(1,1,3)$

**Problem 1.61**   Find the tangent plane at $(0,0)$ to

$$g(x, y) = e^{-(x^2+y^2)}$$

**Problem 1.62**   Find the tangent plane at $(\pi/4, \pi/4, \pi/4)$ to

$$\cos(xyz) = 0$$

**Problem 1.63**   If

$$x^2 + y^2 + z^2 = r^2$$

is a sphere of radius $r$ centered at the origin $(0,0,0)$ (it is), show that a sphere has a tangent plane at right angles to any non-zero vector.

**Problem 1.64**   If you want to find the tangent plane to a point on a sphere, what is the simplest method? Explain.

**Problem 1.65**   If $P$ and $Q$ are planes that are both at right angles to a vector $\vec{v}$, and they are not equal, what can be said about the intersection of $P$ and $Q$?

**Problem 1.66**   Suppose that $\vec{u}$ and $\vec{v}$ are vectors so that $\vec{v} \cdot \vec{u} = 0$. If $P$ is a plane at right angles to $\vec{u}$, and $Q$ is a plane at right angles to $\vec{v}$ in three-dimensional space, then what is the most that can be said about the intersection of the planes?

CHAPTER 2

# Multivariate and Constrained Optimization

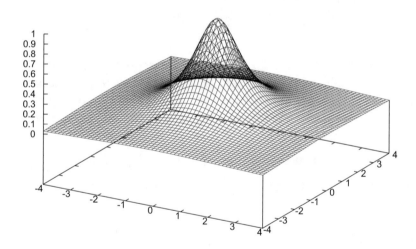

Figure 2.1: A graph of $f(x, y) = \dfrac{1}{x^2 + y^2 + 1}$ on $-4 \le x, y \le 4$.

A big application of derivatives in our earlier work was optimizing functions. In this chapter we learn to do this for multiple independent variables. In some ways we will be doing pretty much the same thing – but as always, more variables makes the process more complicated.

## 2.1  OPTIMIZATION WITH PARTIAL DERIVATIVES

Examine the hill-shaped graph in Figure 2.1. This function has a single optimum at $(x, y) = (0, 0)$. It is also positive everywhere. It is a good, simple function for demonstration purposes. The level curves are circles centered on the origin – except at the origin on top of the optimum, where we get a "level curve" consisting of a single point. This sort of point – where the directional derivatives are all zero – is the three-dimensional equivalent of a critical point.

### Knowledge Box 2.1

### Critical points for surfaces

*If $z = f(x, y)$ defines a surface, then the **critical points** of the function are at the points $(x, y)$ that solve the equations:*

$$f_x(x, y) = 0 \text{ and } f_y(x, y) = 0.$$

*As with our original optimization techniques, local and global optima occur at critical points or at the boundaries of the domain of optimization.*

The information in Knowledge Box 2.1 gives us a good start on finding optima of multivariate functions – but it also contains a land mine. The phrase *the boundaries of the domain of optimization* is more fearsome when there are more dimensions. The next section of this chapter is about dealing with some of the kinds of boundaries that arise when optimizing surfaces.

**Example 2.1**   Demonstrate that the function in Figure 2.1, $f(x, y) = \dfrac{1}{x^2 + y^2 + 1}$, in fact has a critical point at (0,0) by solving for the partial derivatives equal to zero.

**Solution:**
The partial derivatives are:

$$f_x(x, y) = \frac{2x}{(x^2 + y^2 + 1)^2} \text{ and } f_y(x, y) = \frac{2y}{(x^2 + y^2 + 1)^2}$$

Remembering that a fraction is zero only when its numerator is zero – and noting the denominators of these partial derivative are never zero – we see we are solving the very difficult system of simultaneous equations:

$$2x = 0 \text{ and } 2y = 0$$

So we verify a single critical point at (0,0).

◊

**Example 2.2** Find the critical point(s) of

$$g(x, y) = x^2 + 2y^2 + 4xy - 6x - 8y + 2$$

**Solution:**

Start by computing the partials.

$$g_x(x, y) = 2x + 4y - 6$$
$$g_y(x, y) = 4y + 4x - 8$$

This gives us the simultaneous system:

$$2x + 4y - 6 = 0$$
$$4x + 4y - 8 = 0$$

$$
\begin{aligned}
2x + 4y &= 6 \\
4x + 4y &= 8 \\
2x &= 2 \qquad &&\text{Second line minus first.} \\
x &= 1 \\
2 + 4y &= 6 \qquad &&\text{Plug x=1 into first line.} \\
4y &= 4 \\
y &= 1
\end{aligned}
$$

So we find a single critical point at $(x, y) = (1, 1)$.

$\Diamond$

Now that we can locate critical points of surfaces, we have to deal with figuring out if the point is a local maximum, a local minimum, or something else. Let's begin by understanding the option of "something else." Examine the function in Figure 2.2. The partial derivatives are $f_x = 2x$ and $f_y = 2y$. So, it has a critical point at (0,0) in the center of the graph, but it *does not* have an optimum. This is a type of critical point called a **saddle point**.

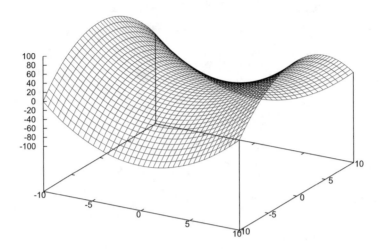

Figure 2.2: A graph of $f(x,y) = x^2 - y^2$ on $-10 \leq x, y \leq 10$. This graph exemplifies a saddle point.

### Knowledge Box 2.2

#### Types of critical points for surfaces

*When we find critical points by solving $f_x = 0$ and $f_y = 0$, there are three possible outcomes:*

- *if all nearby points are lower, the point is a **local maximum**,*

- *if all nearby points are higher, the point is a **local minimum**,*

- *if there are nearby points that are both higher and lower, the point is a **saddle point**.*

If we have a good graph of a function, then we can often look at a critical point to determine its character. This is not always practical. The sign chart technique from *Fast Start Differential Calculus* is not possible – it only makes sense in a one-dimensional setting. There *is* an analog to the second derivative test. Look at the functions in Figures 2.1 and 2.2. If we slice them along the $x$ and $y$ axes, we see that the hill in Figure 2.1 is concave down along both axes, while the saddle in Figure 2.2 is concave down along the $y$ axis and concave up along the $x$ axis. This type of observation can be generalized into a type of second derivative test.

**Knowledge Box 2.3**

### Classifying critical points for surfaces

*For a function $z = f(x, y)$ that defines a surface, define*

$$D(x, y) = f_{xx}(x, y) f_{yy}(x, y) - f_{xy}^2(x, y).$$

*The quantity $D$ is called the **discriminant** of the system. Then for a critical point $(a, b)$*

- *If $D(a, b) > 0$ and $f_{xx}(a, b) > 0$ or $f_{yy}(a, b) > 0$, then the point is a local minimum.*

- *If $D(a, b) > 0$ and $f_{xx}(a, b) < 0$ or $f_{yy}(a, b) < 0$, then the point is a local maximum.*

- *If $D(a, b) < 0$, then the point is a saddle point.*

- *If $D(a, b) = 0$, then the test yields no information about the type of the critical point.*

The fourth outcome – no information – forces you to examine the function in other ways. This situation is usually the result of having functions with repeated roots or high powers of both variables. The function

$$q(x, y) = x^3 + y^3,$$

for example, has a critical point at $(0, 0)$, but all of $f_{xx}$, $f_{yy}$, and $f_{x,y}$ are zero. The critical point is a saddle point, but that conclusion follows from careful examination of the graph of the function, not from the second derivative test.

A part of the structure of this multivariate version of the second derivative test is that *either $f_{xx}$ or $f_{yy}$* may be used to check if a critical point is a minimum or maximum. A frequent question by students is "which should I use?" The answer is "whichever is easier." The reason for this is that the outcome of the discriminant test that tells you a critical point is an optimum of some sort *forces* the signs of $f_{xx}$ and $f_{yy}$ to agree.

Examine the formula for $D(a, b)$ and see if you can tell why this is true.

**Example 2.3**   Find and classify the critical point of

$$h(x, y) = 4 + 3x + 5y - 3x^2 + xy - 2y^2$$

**Solution:**

The partials are $f_x = 3 - 6x + y$ and $f_y = 5 + x - 4y$. Setting these equal to zero we obtain the simultaneous system of equations:

$$6x - y = 3$$
$$-x + 4y = 5$$

$$\text{Solve:}$$
$$-6x + 24y = 30$$

$$23y = 33$$

$$y = 33/23$$

$$-x + 132/23 = 115/23$$

$$-x = -17/23$$

$$x = 17/23$$

Now compute $D(x, y)$. We see $f_{xx} = -6$, $f_{yy} = -4$ and $f_{xy} = 1$. So,

$$D = (-6)(-4) - 1^2 = 23.$$

Since $D > 0$, the critical point is an optimum of some sort. The fact that $f_{xx} < 0$ or the fact that $f_{yy} < 0$ suffices to tell us this point is a maximum.

◊

Notice that in Example 2.3 the equation is a quadratic with large negative squared terms and a small mixed term $(xy)$. The fact that its sole critical point is a maximum means that it is a paraboloid opening downward. In the next example we look at a quadratic with a large mixed term.

**Example 2.4**    Find and classify the critical points of

$$q(x, y) = x^2 + 6xy + y^2 - 14x - 10y + 3.$$

**Solution:**

Start by computing the needed partials and the discriminant.

$$f_x(x, y) = 2x + 6y - 14$$

$$f_y(x, y) = 6x + 2y - 10$$

$$f_{xx} = 2$$

$$f_{yy} = 2$$

$$f_{xy} = 6$$

$$D(x, y) = 2 \cdot 2 - 6^2 = -32$$

These calculations show that any critical point we find will be a saddle point since $D < 0$. To find the critical point we solve the first partials equal to zero obtaining the system

$$2x + 6y = 14$$

$$6x + 2y = 10$$

$$6x + 18y = 42 \qquad\qquad\qquad\qquad \text{3x line one}$$

$$16y = 32 \qquad\qquad\qquad\qquad \text{Line 3 minus line two}$$

$$y = 2$$

$$2x + 12 = 14 \qquad\qquad\qquad\qquad \text{Substitute}$$

$$2x = 2$$

$$x = 1$$

So we see that the function has, as its sole critical point, a saddle point at (1,2).

◊

The last two examples have been bivariate quadratic equations with a constant discriminant. This is a feature of all bivariate quadratics. All functions of this kind have a single critical point and a constant discriminant. For the next example we will tackle a more challenging example.

**Example 2.5** Find and classify the critical points of $f(x, y) = x^4 + y^4 - 16xy + 6$.

**Solution:**

Start by computing the needed partials and the discriminant.

$$f_x(x, y) = 4x^3 - 16y$$

$$f_y(x, y) = 4y^3 - 16x$$

$$f_{xx}(x, y) = 12x^2$$

$$f_{yy}(x, y) = 12y^2$$

$$f_{xy}(x, y) = -16$$

$$D(x, y) = f_{xx}f_{yy} - f_{xy}^2$$

$$D(x, y) = 144x^2y^2 - 256$$

Next, we need to find the critical points by solving the system

$$4x^3 = 16y$$

$$4y^3 = 16x$$

or

$$x^3 = 4y$$

$$y^3 = 4x$$

Solve the second equation to get $y = \sqrt[3]{4x}$ and plug into the first, obtaining:

$$x^3 = 4\sqrt[3]{4x}$$

$$x^9 = 256x \qquad\qquad \text{Cube both sides}$$

$$x^9 - 256x = 0$$

$$x(x^8 - 256) = 0$$

$$x = 0, \pm \sqrt[8]{256}$$

$$x = 0, \pm 2$$

Since the equation system is symmetric in $x$ and $y$ we may deduce that $y = 0, \pm 2$ as well. Referring back to the original equations, it is not hard to see that, when $x = 0$, $y = 0$; when $x = 2$ so must $y$; when $x = -2$ so must y. This gives us three critical points: $(-2, -2), (0, 0)$, and $(2, 2)$. Next, we check the discriminant at the critical points.

$$D(-2, -2) = 2304 - 256 = 2048 > 0$$

Since $f_{xx}(-2, -2) > 0$, this point is a minimum.

$$D(2, 2) = 2048 > 0$$

Another minimum, and

$$D(0, 0) = -256,$$

so this point is a saddle point.

$$\diamondsuit$$

One application of multivariate optimization is to minimize distances. For that we should state, or re-state, the definition of distance.

### Knowledge Box 2.4

#### The definition of distance

If $p = (x_0, y_0)$ and $q = (x_1, y_1)$ are points in the plane, then the distance between $p$ and $q$ is

$$d(p, q) = \sqrt{(x_0 - x_1)^2 + (y_0 - y_1)^2}.$$

If $r = (x_0, y_0, z_0)$ and $s = (x_1, y_1, z_1)$ are points in space, then the distance between $r$ and $s$ is

$$d(r, s) = \sqrt{(x_0 - x_1)^2 + (y_0 - y_1)^2 + (z_0 - z_1)^2}.$$

The definition in Knowledge Box 2.4 can be extended to any number of dimensions – the distance between two points is the square root of the sum of the squared differences of the individual coordinates of the point. For this text only two- and three-dimensional distances are required. With the definition of distance in place, we can now pose a standard type of problem.

**Example 2.6**  What point on the plane $z = 3x + 2y + 4$ is closest to the origin?

**Solution:**

The origin is the point $(0, 0, 0)$, while a general point $(x, y, z)$ on the plane has the form $(x, y, 3x + 2y + 4)$. That means that the distance we are trying to minimize is given by

$$d = \sqrt{(x - 0)^2 + (y - 0)^2 + (3x + 2y + 4 - 0)^2}$$

$$= \sqrt{10x^2 + 12xy + 5y^2 + 24x + 16y + 16}$$

The partial derivatives of this function are:

$$d_x = \frac{20x + 12y + 24}{2\sqrt{10x^2 + 12xy + 5y^2 + 24x + 16y + 16}}$$

$$d_y = \frac{12x + 10y + 16}{2\sqrt{10x^2 + 12xy + 5y^2 + 24x + 16y + 16}}$$

We need to solve the simultaneous equation in which each of these partials is zero. Remember a fraction is zero only where its numerator is zero. This maxim gives us the simultaneous equations:

$$20x + 12y + 24 = 0$$

$$12x + 10y + 16 = 0$$

Simplify

$$5x + 3y = -6$$

$$6x + 5y = -8$$

Solve:

$$30x + 18y = -36$$

$$30x + 25y = -40$$

$$7y = -4$$

$$y = -4/7$$

$$5x - 12/7 = -42/7$$

$$5x = -30/7$$

$$x = -6/7$$

So we find a single critical point $(-6/7, -4/7)$ Considering the geometry of a plane, we see that it has a unique closest approach to the origin. So, computing $z = 3x + 2y + 4 = -18/7 - 8/7 + 28/7 = 2/7$, gives us that the point on the plane closest to the origin is

$$(-6/7, -4/7, 2/7).$$

◊

Clever students will have noticed that when we were minimizing

$$\sqrt{10x^2 + 12xy + 5y^2 + 24x + 128 + 16}$$

the numerators of the partial derivatives were exactly the *partial derivatives* of $10x^2 + 12xy + 5y^2 + 24x + 128 + 16$. This observation is an instance of a more general shortcut.

### Knowledge Box 2.5

### Minimization of distance—a shortcut

*Suppose that we are optimizing a function $g(x, y) = \sqrt{f(x, y)}$ or $g(x, y, z) = \sqrt{f(x, y, z)}$. Then the critical points for optimization of $g$ and $f$ are the same, and the second derivative tests agree. The actual values of the function are not, which means care is required.*

Why does this shortcut work? It is because if $0 < a < b$, then $0 < \sqrt{a} < \sqrt{b}$. So optimizing the value finds where the optimum of the square root of the value is as well.

**Example 2.7**  Minimize:

$$g(x, y) = \sqrt{x^2 + y^2 + 3}$$

**Solution:**

In this case $g(x, y) = \sqrt{f(x, y)}$ where $f(x, y) = x^2 + y^2 + 3$, so the shortcut applies. Finding the relevant partials we see that

$$f_x(x, y) = 2x$$
$$f_y(x, y) = 2y$$

which is easy to see has a critical point at (0,0). The second derivative test shows that:

$$f_{xx} f_{yy} - f_{xy}^2 = 2 \cdot 2 - 0 = 4 > 0$$

So the critical point is an optimum. Since $f_{xx} = 2 > 0$, it is a minimum. This means the minimum value of $g(x, y)$ is at the point $(0, 0, \sqrt{3})$.

$$\diamond$$

**Definition 2.1** *A function $m(x)$ is* **monotone increasing** *if, whenever $a < b$ and $m(x)$ exists on $[a, b]$, then $m(a) < m(b)$.*

Notice that $m(x) = \sqrt{x}$ is monotone increasing. In fact the shortcut in Knowledge Box 2.5 works for any monotone increasing function. These functions include: $e^x$, $\ln(x)$, $\tan^{-1}(x)$, $x^n$ when $n$ is odd, and $\sqrt[n]{x}$.

### Knowledge Box 2.6

### A test for a function being monotone increasing

*We know that a function is increasing if its first derivative is positive. If a function exists and has a positive derivative on the interval [a,b] it is a monotone increasing function on [a,b].*

**Example 2.8**    Show that $y = \ln(x)$ is monotone increasing where it exists.

**Solution:**

According to Knowledge Box 2.6, a function is increasing when its first derivative is positive. The function $y = \ln(x)$ only exists on the interval $(0, \infty)$. Its derivative is $y' = 1/x$ which is positive for any positive $x$. Thus, $\ln(x)$ is increasing everywhere that it exists and so is a monotone increasing function.

$$\Diamond$$

Using an extension of the shortcut from Knowledge Box 2.5 with these other monotone functions will make the homework problems much easier. Let's practice.

**Example 2.9**    Find the point of closest approach to the origin on the plane

$$3x - 4y - z = 4.$$

**Solution:**

First find the distance between the origin $(0, 0, 0)$ and a generic point $(x, y, 3x - 4y - 4)$ on the plane. We get that

$$d(x, y) = \sqrt{x^2 + y^2 + 9x^2 - 24xy + 16y^2 - 32y - 24x + 16}$$

$$d(x, y) = \sqrt{10x^2 - 24xy + 17y^2 - 32y - 24x + 16}$$

Using the distance minimization shortcut, find the critical points of

$$d^2(x, y) = 10x^2 - 24xy + 17y^2 - 32y - 24x + 16$$

which is the relatively simple quadratic case.

$$f_x(x, y) = 20x - 24y - 24$$

$$f_y(x, y) = -24x + 34y - 32$$

$$f_{xx}(x, y) = 20$$

$$f_{yy}(x, y) = 34$$

$$f_{x,y}(x, y) = -24$$

$$D(x, y) = 20 \cdot 34 - (-24)^2 = 104$$

So we see that $D > 0$ and $f_{xx} > 0$ meaning we will find a minimum distance. Now we must solve for the critical point.

$$20x - 24y = 24$$

$$-24x + 34y = 32$$

$$5x - 6y = 6$$

$$-12x + 17y = 16$$

$$60x - 72y = 72$$

$$-60x + 85y = 80$$

$$13y = 152$$

$$y = 152/13$$

$$5x - 912/13 = 78/13$$

$$5x = 990/13$$

$$x = 198/13$$

So the critical point is roughly at $x = 15.23$, $y = 11.69$, making the point of closest approach $(15.23, 11.69, 2.93)$. Not a lot easier – but we avoided all sorts of square roots.

◊

# PROBLEMS

**Problem 2.10**   Find the critical points for each of the following functions, and use the second derivative test to classify them as maxima, minima, or saddle points. Also, find the value of the function at the critical point.

1. $f(x, y) = x^2 + y^2 - 6x - 4y + 4$

2. $g(x, y) = 3x^2 y + y^3 - 3x^2 - 3y^2 + 2$

3. $h(x, y) = 12 - x^2 + 3xy - y^2 + 2x - y + 1$

4. $q(x, y) = x^2 + xy + y^2 + 4x - 5y + 2$

5. $r(x, y) = x^4 + y^4 - 4xy + 1$

6. $s(x, y) = x^3 + 2y^3 - 4xy$

**Problem 2.11**   Show that

$$f(x, y) = x^4 + y^4$$

is an example of a function where the second derivative test yields no useful information.

**Problem 2.12**   Suppose that

$$f(x, y) = ax^2 + bxy + cy^2 + dx + ey + f.$$

Show that this function has at most one critical point. When it does have a critical point, derive rules based on the constants $a$-$f$ for classifying that critical point.

**Problem 2.13**   Find the point on the plane

$$x + y + z = 4$$

closest to the origin.

**Problem 2.14**   Find the point on the plane

$$z = -x/3 + y/4 + 1$$

closest to the origin.

**Problem 2.15**   Find the point on the tangent plane of

$$f(x, y) = x^2 + y^2 + 1$$

at $(2, 1, 6)$ that is closest to the origin.

**Problem 2.16**   Find the critical points for each of the following functions, and use the second derivative test to classify them as maxima, minima, or saddle points. Also, find the value of the function at the critical point.

1. $f(x, y) = \ln(x^2 + y^2 + 4)$

2. $g(x, y) = e^{x^4 + y^4 - 36xy + 6}$

3. $h(x, y) = \tan^{-1}(x^3 + 2y^3 - 4xy)$

4. $r(x, y) = (x^2 + xy + 3y^2 - 4x + 3y + 1)^5$

5. $s(x, y) = \sqrt{x^2 + y^2 + 14}$

6. $q(x, y) = \tan^{-1}\left(e^{x^2 + xy + y^2 - 2x - 3y + 1}\right)$

**Problem 2.17**    For each of the following functions either demonstrate it is not monotone increasing on the given interval or show that it is.

1. $y = e^x$ on $(-\infty, \infty)$

2. $y = x^3 - x$ on $(-\infty, \infty)$

3. $y = \dfrac{x}{x^2 + 1}$ on $(-1, 1)$

4. $y = \tan^{-1}(x)$ on $(-\infty, \infty)$

5. $y = \dfrac{x^2}{x^2 + 1}$ on $(0, \infty)$

6. $y = \dfrac{2x}{x^2 + 1}$ on $(-\infty, \infty)$

7. $y = \dfrac{e^x}{e^x + 1}$ on $(-\infty, \infty)$

**Problem 2.18**    Suppose that we are going to cut a line of length $L$ into three pieces. Find the division into pieces that maximizes the sum of the squares of the lengths. This can be phrased as a multivariate optimization problem.

**Problem 2.19**    Find the largest interval on which the function

$$\frac{x}{x^2 + 16}$$

is monotone increasing.

**Problem 2.20**    Find the largest interval on which the function

$$\frac{x^2 - 1}{x^2 + 4}$$

is monotone increasing.

## 2.2    THE EXTREME VALUE THEOREM REDUX

The extreme value theorem, last seen in *Fast Start Differential Calculus*, says that *The global maximum and minimum of a continuous, differentiable function must occur at critical points or at the boundaries of the domain where optimization is taking place.* This is just as true for optimizing a function $z = f(x, y)$. So what's changed? The largest change is that the boundary can now have a very complex shape.

In the initial section of this chapter we carefully avoided optimizing functions with boundaries, thus avoiding the issue with boundaries. In the next couple of examples we will demonstrate techniques for dealing with boundaries.

**Example 2.21**   Find the global maximum of

$$f(x, y) = (xy + 1)e^{-x-y}$$

for $x, y \geq 0$.

**Solution:**

This function is to be optimized only over the first quadrant. We begin by finding critical points in the usual fashion

$$f_x(x, y) = ye^{-x-y} + (xy + 1)e^{-x-y}(-1) = (-xy + y - 1)e^{-x-y}$$

$$f_y(x, y) = xe^{-x-y} + (xy + 1)e^{-x-y}(-1) = (-xy + x - 1)e^{-x-y}$$

Since the extreme value theorem tells us that the optima occur at boundaries or critical points, we won't need a second derivative test – we just compare values. This means out next step is to solve for any critical points. Remember that powers of e cannot be zero, giving us the system of equations:

$$-xy + y - 1 = 0$$

$$-xy + x - 1 = 0$$

$$xy = x - 1 = y - 1 \qquad\qquad\qquad \text{so } x = y$$

$$-x^2 + x - 1 = 0$$

$$x^2 - x + 1 = 0$$

$$x = \frac{1 \pm \sqrt{1 - 4}}{2} \qquad\qquad \text{There are no critical points!}$$

This means that the extreme value occurs on the boundaries – the positive $x$ and $y$ axes where either $x = 0$ or $y = 0$. So, we need the largest value of $f(0, y) = e^{-y}$ for $y \geq 0$ and $f(x, 0) = e^{-x}$ for $x \geq 0$. Since $e^{-x}$ is largest (for non-negative $x$) at $x = 0$, we get that the global maximum is $f(0, 0) = 1$.

◊

While the extreme value theorem lets us make decisions without the second derivative test, it forces us to examine the boundaries, which can be hard. It is sometimes possible to solve the problem by adopting a different point of view. The next example will use a transformation to a parametric curve to solve the problem.

**Example 2.22**   Maximize

$$g(x, y) = x^4 + y^4$$

for those points $\{(x, y) : x^2 + y^2 \le 4\}$.

**Solution:**

This curve has a single critical point at $(x, y) = (0, 0)$ which is extremely easy to find. The boundary for the optimization domain is the circle $x^2 + y^2 = 4$, a circle of radius 2 centered at the origin. This boundary is also the parametric curve $(2\cos(t), 2\sin(t))$. This means that *on the boundary* the function is

$$g(2\cos(t), 2\sin(t)) = 16\cos^4(t) + 16\sin^4(t).$$

This means we can treat the location of optima on the boundary as a single-variable optimization task of the function $g(t) = 16\cos^4(t) + 16\sin^4(t)$. We see that

$$g'(t) = 64\cos^3(t) \cdot (-\sin(t)) + 64\sin^3(t)\cos(t) = 64\sin(t)\cos(t)(\sin^2(t) - \cos^2(t))$$

Solve:

$$\sin(t) = 0 \qquad\qquad\qquad t = (2n + 1)\frac{\pi}{2}$$

$$\cos(t) = 0 \qquad\qquad\qquad t = n\pi$$

$$\sin^2(t) - \cos^2(t) = 0$$

$$\sin^2(t) = \cos^2(t) \qquad\qquad\qquad t = \pm(2n + 1)\frac{\pi}{4}$$

If we plug these values of $t$ back into the parametric curve, the points on the boundary that may be optima are: $(0, \pm 2)$, $(\pm 2, 0)$, and $(\pm\sqrt{2}, \pm\sqrt{2})$. Plugging the first four of these into the curve we get that $g(x, y) = 16$. Plugging the last four into the function we get $g(x, y) = 2 \cdot (\sqrt{2})^4 = 2 \cdot 4 = 8$. At the critical point $(0,0)$ we see $g(x, y) = 0$. This means that the maximum value of $g(x, y)$ on the optimization domain is 16 at any of $(0, \pm 2)$, $(\pm 2, 0)$.

◊

Notice that the solutions to the parametric version of the boundary gave us an infinite number of solutions. But, when we returned to the $(x, y, z)$ domain, this infinite collection of solutions were just the eight candidate points repeated over and over. This means that our failure to put a bound on the parameter $t$ was not a problem; bounding $t$ was not necessary.

A problem with the techniques developed in this section is that all of them are special purpose. We will develop general purpose techniques in Section 2.3 – the understanding of which is substantially aided by the practice we got in this section.

**Example 2.23**   If

$$h(x, y) = x^2 + y^2$$

find the minimum value of $h(x, y)$ among those points $(x, y)$ on the line $y = 2x - 4$.

**Solution:**

The function $h(x, y)$ is a paraboloid that opens upward. If we look at the points $(x, y, z)$ on $h(x, y)$ that happen to lie on a line, that line will slice a parabolic shape out of the surface defined by $h(x, y)$. First note that $h(x, y)$ has a single critical point at $(0, 0)$ which is *not in* the domain of optimization. This means we may consider only those points on $y = 2x - 4$, in other words on the function

$$h(x, 2x - 4) = x^2 + (2x - 4)^2 = 5x^2 - 16x + 16$$

This means that the problem consists of finding the minimum of $f(x) = 5x^2 - 16x + 16$.

$$f'(x) = 10x - 16 = 0$$

$$10x = 16$$

$$x = 8/5$$

$$y = 16/5 - 20/5 = -4/5$$

Since $f(x)$ opens upward, we see that this point is a minimum; there are no boundaries to the line that is constraining the values of $(x, y)$ so the minimum value of $h(x, y)$ on the line is

$$h(8/5, -4/5) = \frac{64}{25} + \frac{16}{25} = \frac{80}{25} = \frac{16}{5}$$

◇

Example 2.23 did not really use the extreme value theorem. For that to happen we would need to optimize over a line segment instead of a full line.

**Example 2.24**  If

$$h(x, y) = x^2 + y^2$$

find the minimum and maximum value of $h(x, y)$ among those points $(x, y)$ on the line $y = x + 1$ for $-4 \le x \le 4$.

**Solution:**

This problem is very similar to Example 2.23. The ends of the line segment are (-4,-3) and (4,5), found by substituting into the formula for the line. On the line $h(x, y)$ becomes

$$h(x, x + 1) = x^2 + (x + 1)^2 = 2x^2 + 2x + 1$$

So we get a critical point at $4x + 2 = 0$ or $x = -1/2$ which is the point $(-1/2, 1/2)$.

Plug in and we get

$$h(-4, -3) = 25$$

$$h(-1/2, 1/2) = 1/2$$

$$h(4, 5) = 41$$

This means the minimum value is 1/2 at the critical point and that the maximum value is 41 at one of the endpoints of the domain of optimization.

◇

Making the domain a line segment is one of the simplest possible options. Let's look at another example with a more complex domain of optimization.

**Example 2.25**  If

$$s(x, y) = 2x + y - 4$$

find the minimum and maximum value of $s(x, y)$ among those points $(x, y)$ on the curve $y = x^2 - 3$ for $-2 \le x \le 2$.

**Solution:**

In this problem we are cutting a parabolic segment out of the plane $s(x, y) = 2x + y - 4$. We get that the ends of the domain of optimization are $(2, 1)$ and $(-2, 1)$ by plugging in the ends of the interval in $x$ to the formula for the parabolic segment. Substituting the parabolic segment into the plane yields

$$s(x, x^2 - 3) = 2x + x^2 - 3 - 4 = x^2 + 2x - 7$$

This means our critical point appears at $2x + 2 = 0$ or $x = -1$, making the candidate point $(-1, -2)$. Plugging the candidate points into $s(x, y)$ yields:

$$s(-2, 1) = -7$$

$$s(-1, -2) = -8$$

$$s(2, 1) = 1$$

This means that the maximum value is 1 at $(2, 1)$, and the minimum is $-8$ at $(-1, -2)$.

$$\diamond$$

Some sets of boundaries are simple enough that we can use geometric reasoning to avoid needing to use calculus on the boundaries.

**Example 2.26**    Find the maximum value of

$$h(x, y) = x^2 + y^2$$

for $-2 \le x, y \le 3$.

**Solution:**

First of all, we know that this surface has a single critical point at $(0, 0)$ – this surface is an old friend (see graph in Figure 1.2).

The domain of optimization is a square with corners $(-2, 2)$ and $(3, 3)$. Along each side of the square, we see that the boundary is a line – and so has extreme values at its ends. This means that we need only check the corners of the square; the interior of the edges – viewed as line segments – cannot attain maximum or minimum values *by the extreme value theorem*. This means we need only add the points $(-2, -2)$, $(-2, 3)$, $(3, -2)$, and $(3, 3)$ to our candidates.

Plugging in the candidate points we obtain:

$$h(-2-2) = 8$$

$$h(0,0) = 0$$

$$h(-2,3) = 13$$

$$h(3,-2) = 13$$

$$h(3,3) = 18$$

So the maximum is 18 at $(3,3)$, and the minimum is 0 at $(0,0)$.

$$\Diamond$$

The goal for this section was to set up LaGrange Multipliers – the topic of Section 2.3. The take-home message from this section is that the extreme value theorem implies that optimizing on a boundary is the difficult added portion of optimizing on a bounded domain.

The techniques that we will develop in the next section require that the boundary itself be a differentiable curve. Some of the boundaries in this section are made of several differentiable curves, meaning that the techniques in this section may be easier for those problems. If we have a boundary that is not differentiable, then the techniques in this section are the only option.

## PROBLEMS

**Problem 2.27**    If we look at the points on

$$h(x, y) = x^2 + y^2 + 4x + 4$$

that fall on a line, then there is a minimum somewhere on the line. Find that minimum value for the following lines.

1. $y = 3x + 1$

2. $y = 4 - x$

3. $x + y = 6$

4. $2x - 7y = 24$

5. $x + y = \sqrt{3}/2$

6. $-3x + 5x = 7$

**Problem 2.28**   Find the maximum of

$$f(x, y) = (x^2 + y^2) e^{-xy}$$

with $(x, y)$ in the first quadrant where $0 < x, y$.

Hint: this function is symmetric. Use this fact.

**Problem 2.29**   Find the maximum and minimum of

$$q(x, y) = \frac{1}{x^2 + y^2 - 2x - 4y + 6}$$

in the first quadrant: $0 < x, y$.

**Problem 2.30**   Find the maximum of

$$g(x, y) = (x^4 + y^4)$$

on the set of points

$$\{(x, y) : x^2 + y^2 \le 25\}.$$

**Problem 2.31**   Maximize

$$h(x) = x^4 + y^4$$

on the set of points $\{(x, y) : x^2 + y^2 \le 25\}$.

**Problem 2.32**   If

$$g(x, y) = 2x^2 + 3y^2,$$

find the minimum of $g(x, y)$ for those points $(x, y)$ on the line $y = x - 1$.

**Problem 2.33**   For the function

$$s(x, y) = x^2 + y^2 - 2x + 4y + 1$$

with $(x, y)$ on the following line segments, find the minimum and maximum values.

1. $y = 2x - 2$ on $-2 \leq x \leq 2$

2. $y = x + 7$ on $-1 \leq x \leq 4$

3. $y = 6 - 5x$ on $0 \leq x \leq 6$

4. $x + y = 10$ on $4 \leq x \leq 10$

5. $2x - 7y = 8$ on $-10 \leq x \leq 10$

6. $2x - y = 13$ on $1 \leq x \leq 5$

**Problem 2.34**   Suppose that

$$p(x, y) = 3x - 5y + 2$$

Find the maximum and minimum values on the following parametric curves.

1. $(3t + 1, 5 - t)$;    $-5 \leq t \leq 5$

2. $(\cos(t), \sin(t))$

3. $(\sin(t), 3\cos(t))$

4. $(\sin(2t), \cos(t))$

5. $(t\cos(t), t\sin(t))$;    $0 \leq t \leq 2\pi$

6. $(3\sin(t), 2\sin(t))$

**Problem 2.35**   Find the maximum and minimum of

$$q(x, y) = \frac{1}{x^2 + y^2 + 4x - 12y + 45}$$

in the first quadrant: $0 < x, y$.

**Problem 2.36**   Suppose that $P = f(x, y)$ is a plane and that we are considering the points where $x^2 + y^2 = r^2$ for some constant $r$. If $f(x, y)$ is not equal to a constant, explain why there is a unique minimum and a unique maximum value.

**Problem 2.37**   If

$$f(x, y) = x^2 + y^2,$$

find the minimum of $f(x, y)$ for those points $(x, y)$ on the line $y = mx + b$.

# 2.3  LAGRANGE MULTIPLIERS

This section introduces **constrained optimization** using a technique called **Lagrange multipliers**. The basic idea is that we want to optimize a function $f(x, y)$ at those points where $g(x, y) = c$. The function $g(x, y)$ is called the **constraint**.

The proof that Lagrange multipliers work is beyond the scope of this text, so we will begin by just stating the technique.

**Knowledge Box 2.7**

> ### The method of LaGrange Multipliers with two variables
>
> *Suppose that $z = f(x, y)$ defines a surface and that $g(x, y) = c$ specifies points of interest. Then the optima of $f(x, y)$, subject to the constraint that $g(x, y) = c$, occur at solutions to the system of equations*
>
> $$f_x(x, y) = \lambda \cdot g_x(x, y)$$
>
> $$f_y(x, y) = \lambda \cdot g_y(x, y)$$
>
> $$g(x, y) = c$$
>
> *where $\lambda$ is an* **auxiliary variable**.

The variable $\lambda$ is new and strange – it is the "multiplier" – and, as we will see, correct solutions to the system of equations that arise from Lagrange multipliers typically use $\lambda$ in a fashion that causes it to drop out.

If the constraint $g(x, y) = c$ is thought of as the boundary of the domain of optimization, then using Lagrange multipliers gives us a tool for resolving the boundary as a source of optima as per the extreme value theorem.

With that context, let's practice our Lagrange multipliers.

**Example 2.38**    Find the maxima and minima of

$$f(x, y) = x + 4y - 2$$

on the ellipse $2x^2 + 3y^2 = 36$.

**Solution:**

The equations arising from the Lagrange multiplier technique are:

$$1 = \lambda \cdot 4x$$

$$4 = \lambda \cdot 6y$$

$$2x^2 + 3y^2 = 36$$

Solving:

$$x = \frac{1}{4\lambda}$$

$$y = \frac{2}{3\lambda}$$

Plug into the constraint and we get:

$$\frac{2}{16\lambda^2} + \frac{12}{9 \cdot \lambda^2} = 36$$

$$\frac{1}{8} + \frac{4}{3} = 36\lambda^2$$

$$35/24 = 36\lambda^2$$

$$35/864 = \lambda^2$$

or

$$\lambda = \pm\sqrt{35/864}$$

Which yields candidate points:

$$x = \pm \frac{1}{4}\sqrt{\frac{864}{35}} = \pm\sqrt{\frac{54}{35}}$$

$$y = \pm \frac{2}{3}\sqrt{\frac{864}{35}} = \pm\sqrt{\frac{384}{35}}$$

Since the equation of $f(x, y)$ is a plane that gets larger as $x$ and $y$ get larger, we see that the maximum is

$$f\left(\sqrt{54/35}, \sqrt{384/35}\right) \cong 12.5$$

and the minimum is

$$f\left(-\sqrt{54/35}, -\sqrt{384/35}\right) \cong -16.5$$

◊

Notice that in the calculations in Example 2.38, the auxiliary variable $\lambda$ served as an informational conduit that let us discover the candidate points. The next example is much simpler.

**Example 2.39**   Find the point of closest approach of the line $2x + 5y = 3$ to the origin (0,0).

**Solution:**

This problem is like the closest approach of a plane to the origin problems, but in a lower dimension. The tricky part of this problem is phrasing it as a function to be minimized and a constraint.

Since we are minimizing distance we get that the function is the distance of a point $(x, y)$ from the origin:

$$d(x, y) = \sqrt{(x - 0)^2 - (y - 0)^2} = \sqrt{x^2 + y^2}$$

As per the monotone function shortcut, we can instead optimize the function

$$d^2(x, y) = x^2 + y^2.$$

The constraint function is the line. We can phrase the line as a constraint by saying:

$$g(x, y) = 2x + 5y = 3$$

With the parts in place, we can extract the Lagrange multiplier equations.

$$2x = \lambda 2$$

$$2y = \lambda 5$$

$$2x + 5y = 3$$

Solve:

$$x = \lambda$$

$$y = \frac{5}{2} \cdot \lambda$$

$$2(\lambda) + 5\left(\frac{5}{2} \cdot \lambda\right) = 3$$

$$\frac{29}{2} \cdot \lambda = 3$$

$$\lambda = \frac{6}{29}$$

Yielding: $gx = \frac{6}{29}$ and $y = \frac{15}{29}$

This makes the point on $2x + 5y = 3$ closest to the origin $\left(\frac{6}{29}, \frac{15}{29}\right)$.

**Example 2.40**    Find the maximum value of

$$f(x, y) = x^2 + y^3$$

subject to the constraint $x^2 + y^2 = 9$.

**Solution:**

This problem is already in the correct form for Lagrange multipliers, so we may immediately derive the system of equations.

$$2x = 2x\lambda$$

$$2y = 3y^2\lambda$$

$$x^2 + y^2 = 9$$

The first equation yields the useful information that $\lambda = 1$ but that $x$ may take on any value, unless $x = 0$ in which case $\lambda$ may be anything.

Given that $\lambda = 1$, the second equation tells us $y = 0$ or $y = 2/3$. If $\lambda$ is free and $x = 0$ then $y$ may be anything.

Plugging the values for $y$ into the constraint that says the points lie on a circle, we can retrieve values for $x$: when $y = 0$, $x = \pm 3$; when $y = 2/3$, $x = \pm\sqrt{9 - 4/9} = \pm\sqrt{77/9}$.

If $x = 0$ and $y$ is free the constraint yields $y = \pm 3$.

This gives us the candidate points $(\pm 3, 0)$, $(0, \pm 3)$, $(2/3, \sqrt{77}/3)$, and $(2/3, -\sqrt{77}/3)$.

The sign of the $x$-coordinate is unimportant because $f(x, y)$ depends on $x^2$; since $f(x, y)$ depends on $y^3$, positive values yield larger values of $f$, and negative ones yield smaller values of $f$.

Since $f(3, 0) = 9$, $f(2/3, \sqrt{77}/3) \cong 25.47$, and $f(0, 3) = 27$, we get that the maximum value of the function on the circle is:

$$f(0, 3) = 27$$

◇

In Section 2.2 we found the minimum of a line on a quadratic surface that opens upward (Example 2.23). The next example lets us try a problem like this using the formalism of Lagrange multipliers.

**Example 2.41**   Find the minimum of $f(x, y) = x^2 - xy + y^2$ on the line $5x + 7y = 18$.

**Solution:**

The constraint is the line − so $g(x, y) = 5x + 7y = 18$. With this detail we can apply the Lagrange multiplier technique:

$$2x - y = 5\lambda$$

$$2y - x = 7\lambda$$

$$5x + 7y = 18$$

Solve the constraint for $y$ and we get $y = \dfrac{18 - 5x}{7}$ so

$$2x - \frac{18 - 5x}{7} = 5\lambda$$

$$14x - 18 + 5x = 35\lambda$$

$$19x - 35\lambda = 18$$

and

$$2\left(\frac{18 - 5x}{7}\right) - x = 7\lambda$$

$$36 - 10x - 7x = 49\lambda$$

$$17x + 49\lambda = -36$$

$$\lambda = -(17x + 36)/49$$

$$19x + 35 \cdot (17x + 36)/49 = 18$$

$$1526x = -378$$

$$x = -189/763 \cong -0.25$$

$$y = 2097/763 \cong 2.75$$

Making the minimum

$$\left(\frac{-189}{763}\right)^2 - \frac{-189 \cdot 2097}{74^2} + \left(\frac{2097}{763}\right)^2 \cong 8.30$$

◊

This problem actually got harder when we used the Lagrange multiplier formalism to solve it. This is another example of how different tools are good for different problems.

At this point, we introduce the 3-space version of Lagrange multipliers. Or, we could say that this is the version of Lagrange multipliers that uses three independent variables. This version of the Lagrange multiplier technique widens the variety of problems we can work with.

### Knowledge Box 2.8

#### The method of LaGrange Multipliers with three variables

*Suppose that $w = f(x, y, z)$ defines a surface and that $g(x, y, z) = c$ specifies points of interest. Then the optima of $f(x, y, z)$, subject to the constraint that $g(x, y, z) = c$, occur at solutions to the system of equations*

$$f_x(x, y, z) = \lambda \cdot g_x(x, y, z)$$

$$f_y(x, y, z) = \lambda \cdot g_y(x, y, z)$$

$$f_z(x, y, z) = \lambda \cdot g_z(x, y, z)$$

$$g(x, y, z) = c$$

*where $\lambda$ is an* auxiliary variable.

**Example 2.42**    Find the point of intersection of the function $w = 2x + y - z$ and the sphere $x^2 + y^2 + z^2 = 8$ that has the largest value of $w$.

**Solution:**

The problem is already in the correct form to apply Lagrange multipliers.

$$2x = 2\lambda$$

$$2y = \lambda$$

$$2z = -\lambda$$

$$x^2 + y^2 + z^2 = 8$$

Solving the first three equations tell us that

$$x = \lambda \qquad y = \frac{1}{2}\lambda \qquad z = -\frac{1}{2}\lambda$$

Plugging these into the last equation tells us that

$$\lambda^2 + (1/4)\lambda^2 + (1/4)\lambda^2 = 8$$

$$(3/2)\lambda^2 = 8$$

$$\lambda^2 = 16/3$$

$$\lambda = \pm 4/\sqrt{3}$$

So we see that: $x = \pm\dfrac{4}{\sqrt{3}} \qquad y = \pm\dfrac{2}{\sqrt{3}} \qquad z = \pm\dfrac{2}{\sqrt{3}}$

Looking at the formula for $w$, we see that $w$ grows as $x$, $y$, and $-z$.

So the point that maximizes $w$ is: $(x, y, z) = \left(\dfrac{4}{\sqrt{3}}, \dfrac{2}{\sqrt{3}}, -\dfrac{2}{\sqrt{3}}\right)$.

◇

**Example 2.43**   Suppose we divide a rope of length 6 m into three pieces. What size of pieces maximizes the product of the lengths?

**Solution:**

Name the length of the pieces $x$, $y$, $z$. That means the function we are maximizing is $f(x, y, z) = xyz$ and the constraint is $x + y + z = 6$.

Having put the problem into the form for Lagrange multipliers, we can move to the system of equations.

$$yz = \lambda$$

$$xz = \lambda$$

$$xy = \lambda$$

$$x + y + z = 6$$

So $x = \lambda/z = y$ and $z = \lambda/x = y$ making $x = y = z$. Since they sum to 6, we see $x = y = z = 2$ is the sole candidate point. Testing constrained points near $(2, 2, 2)$, like $(1.9, 2.1, 2)$ and $(2, 1.95, 2.05)$, shows that this point is a maximum.

# PROBLEMS

**Problem 2.44**    For each of the following sets of functions and constraints, write out but do not solve the Lagrange multiplier equations.

1. $f(x, y) = \cos(x) \sin(y)$ constrained by $x^2 + y^2 = 8$

2. $g(x, y) = xe^y$ constrained by $2x - 3y = 12$

3. $h(x, y) = \dfrac{x^2}{y^2 + 1}$ constrained by $3x^2 + y^2 = 48$

4. $r(x, y) = 2xy$ constrained by $x^2 + y^2 = 8$

5. $s(x, y) = \ln(x^2 + y^2 + 4)$ constrained by $xy = 9$

6. $q(x, y) = \tan^{-1}(xy + 1)$ constrained by $x^2 - y^2 = 4$

**Problem 2.45**   For each of the following lines, find the point on the line closest to the origin using the method of Lagrange multipliers.

1.  $y = 2x + 1$                    3.  $2x + 3y = 17$                    5.  $y = 5x + 25$

2.  $y = 4 - x$                      4.  $x + y = 12$                      6.  $2x + 4y = 8$

**Problem 2.46**   Using Lagrange multipliers, find the maximum and minimum values of

$$f(x, y) = x^4 + y^4$$

subject to the constraint that

$$x^2 + y^2 = 1.$$

**Problem 2.47**   Suppose that

$$z = 3x - 5y + 2$$

For each of the following constraints, find the maximum and minimum values of $z$ subject to the constraint, if any.

1.  $x^2 + y^2 = 4$                              4.  $x^2 - xy + y^2 = 1$

2.  $x^2 + y^2 = 0.25$                          5.  $x^2 - y^2 = 4$

3.  $2x^2 + 6y^2 = 64$                          6.  $xy = 16$

**Problem 2.48**   If

$$f(x, y) = x^3 - y^2$$

and $x^2 + y^2 = 25$, find the maximum and minimum values $f(x, y)$ can take on.

**Problem 2.49**   Use Lagrange multipliers to find the closest approach of the plane

$$f(x, y) = 3x - y + 2$$

to the origin.

**Problem 2.50**    If
$$f(x, y, z) = x^2 + 4y^2 + 9z^2$$
and $x + y + z = 60$, find the values of $x$, $y$, and $z$ that maximize and minimize $f$.

**Problem 2.51**    If
$$h(x, y, z) = x^2 - y^2 + z^2$$
and $x + y + z = 6$, find the values of $x$, $y$, and $z$ that maximize and minimize $h$.

**Problem 2.52**    If
$$q(x, y, z) = x^2 - y^2 - 2z^2$$
and $x + y + z = 300$, find the values of $x$, $y$, and $z$ that maximize and minimize $q$.

**Problem 2.53**    Find the largest point in intersection of
$$x + 2y + 3z = 4$$
and a cylinder of radius 2 centered on the $z$-axis.

# CHAPTER 3

# Advanced Integration

This chapter covers various sorts of integration that compute volumes and areas. The first, volumes of revolution, is a small twist on the integration methods we have already studied. The other techniques involve multivariate integration, which is both newer and more difficult. This latter subject permits us to compute the volume under a surface as we computed the area under the curve earlier.

## 3.1 VOLUMES OF ROTATION

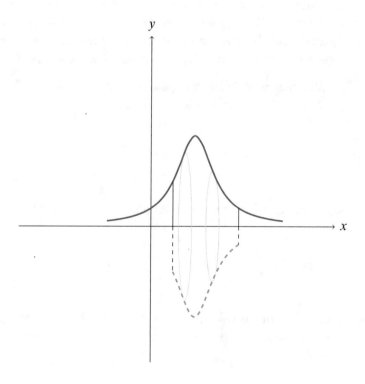

Figure 3.1: A portion of a curve rotated about the $x$-axis.

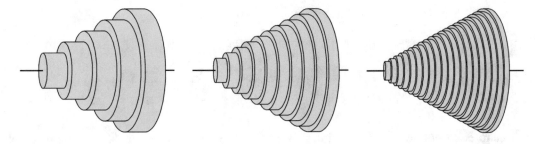

Figure 3.2: Approximations of a rotated volume using smaller and larger numbers of disks.

Figure 3.1 shows a portion of a curve and its shadow reflected about the $x$-axis. This is meant to evoke revolving the curve about the axis between the two vertical lines. When the curve is rotated, it encloses a volume – our goal is to compute that volume.

When integration was defined in *Fast Start Integral Calculus*, we first approximated the area represented by the integral with rectangles and then let the number of rectangles go to infinity, obtaining both the area and the integral as a limit. For volumes of rotation, we use a similar technique with *disks* instead of rectangles. Figure 3.2 shows how a conical volume is approximated using five, ten, and twenty disks. The more disks we use, the closer the sum of the volumes of the disks gets to the volume enclosed by rotating a curve about the $x$ axis.

The final version of the area integral added up functional heights – rectangles of infinite thinness. Recall that the area under a curve from $x = a$ to $x = b$ was

$$\text{Area} = \int_a^b f(x) \cdot dx.$$

Given this, the analogous integral for volume of rotation is

$$\text{Volume} = \int_a^b \text{Area of Circles} \cdot dx = \int_a^b \pi r^2 \cdot dx$$

The radius of the circular slices of the volume we are trying to compute is given by the height of the function being rotated about the axis, giving us the final formula for *volume of rotation of a function about the x-axis*.

**Knowledge Box 3.1**

### Formula for volume of rotation about the $x$ axis

*The volume $V$ enclosed by rotating the function $f(x)$ about the $x$-axis from $x = a$ to $x = b$ is:*

$$V = \pi \int_a^b f(x)^2 \cdot dx.$$

*(Notice that we took the constant $\pi$ out in front of the integral sign.)*

**Example 3.1** Find the volume enclosed by rotating $y = x^2$ about the $x$ axis from $x = 0$ to $x = 2$.

**Solution:**

Start by sketching the situation:

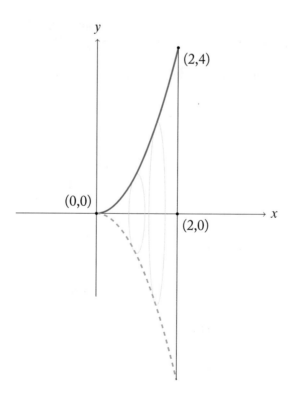

Using the equation in Knowledge Box 3.1 we get:

$$
\begin{aligned}
V &= \pi \int_0^2 f(x)^2 \cdot dx \\[2mm]
&= \pi \int_0^2 (x^2)^2 \cdot dx \\[2mm]
&= \pi \int_0^2 x^4 \cdot dx \\[2mm]
&= \pi \frac{1}{5} x^5 \Big|_0^2 \\[2mm]
&= \pi \left( \frac{32}{5} - 0 \right) \\[2mm]
&= \frac{32\pi}{5} \text{units}^3
\end{aligned}
$$

And so we see the volume of the broad trumpet-shaped solid enclosed by rotating $y = x^2$ about the $x$-axis from $x = 0$ to $x = 2$ has a volume of $32\pi/5 \cong 503$ units$^3$.

$$\diamond$$

The next example is similar except that it uses a more difficult integral. If you're not comfortable with integration by parts, please review it in *Fast Start Integral Calculus*.

**Example 3.2**   Find the volume enclosed by rotating $y = \ln(x)$ about the $x$ axis from $x = 1$ to $x = 4$.

**Solution:**

Start again with a sketch.

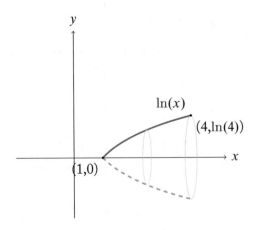

Applying the formula from Knowledge Box 3.1 we get:

$$\text{Volume} = \pi \int_1^4 \ln(x)^2 \cdot dx$$

$$= \pi \left( x \ln(x)^2 - \int_1^4 2 \ln(x) dx \right) \qquad \begin{array}{ll} U = \ln(x)^2 & V = x \\ dU = \dfrac{2\ln(x)}{x} \cdot dx & dV = dx \end{array}$$

$$= \pi \left( x \ln(x)^2 - 2x \ln(x) + 2 \int_1^4 dx \right) \qquad \begin{array}{ll} U = \ln(x) & V = x \\ dU = \dfrac{dx}{x} & dV = dx \end{array}$$

$$= \left. (\pi x \ln(x)^2 - 2\pi x \ln(x) + 2\pi x) \right|_1^4$$

$$= 4\pi \ln(4)^2 - 8\pi \ln(4) + 8\pi - 0 + 0 - 2\pi$$

$$\cong 8.16 \text{ units}^3$$

Now that we have a formula for rotating objects about the $x$-axis, the next logical step is to rotate them about the $y$-axis. This turns out to be a little trickier.

There are two basic techniques.

1. We can figure out a new function (the inverse of the original one) that gives us disks along the $y$-axis, and integrate with respect to $dy$, or

2. We can use a different type of slice, the *cylindrical shell*. Instead of slicing the shape into disks and integrating them, we slice it into cylinders.

The radius of the disks for rotation about the $x$-axis was simply the value of the function, $f(x)$, but the distance from a graph to the $y$ axis is just $x$. We need to get the information about the $y$-distance in somehow. We will start with Method 2, the method that uses a different type of slice (cylinders centered on the $y$-axis). Figure 3.3 shows the result of rotating a line about the $y$ axis.

The surface area of a cylinder, not counting the top and bottom, is $2\pi rh$ where $r$ is its radius and $h$ is its height. The radius, as already noted, is $x$, and the height of the cylinder is $y = f(x)$.

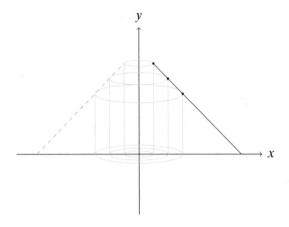

Figure 3.3: Cylindrical shells for computing a volume of revolution about the $y$-axis.

Thus, we have the area of our cylindrical slices

$$\text{Area} = 2\pi \cdot r \cdot h = 2\pi \cdot x \cdot f(x)$$

### Knowledge Box 3.2

**Volume of rotation with cylindrical shells about the $y$-axis**

*If we rotate a function $f(x)$ about the $y$-axis, then the volume enclosed by the curve between $x = a$ and $x = b$ is:*

$$V = 2\pi \int_{x=a}^{x=b} x \cdot f(x) \cdot dx.$$

*(Notice that we took the constant $2\pi$ out in front of the integral sign.)*

One tricky thing about this is that the limits of integration are the range of $x$-values that the radii of the cylindrical shells span. The $y$-distances only come in via the participation of $f(x)$. The formula in Knowledge Box 3.2 assumes we are rotating the area *below the curve* around the $y$ axis. Finding other areas may require a more complicated setup where we need to figure out the height of the cylinders.

**Example 3.3**   Find the volume of rotation of the area below the curve $f(x) = 1/x$ about the $y$-axis from $x = 0.5$ to $x = 2.0$.

**Solution:**

Sketch the situation.

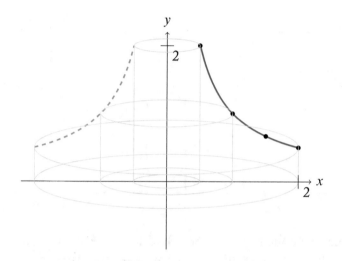

Applying the equation in Knowledge Box 3.2 we get:

$$
\begin{aligned}
\text{Volume} &= 2\pi \int_{0.5}^{2} x \cdot f(x) \cdot dx \\[2mm]
&= 2\pi \int_{0.5}^{2} x \cdot \frac{1}{x} \cdot dx \\[2mm]
&= 2\pi \int_{0.5}^{2} dx \\[2mm]
&= 2\pi x \Big|_{0.5}^{2} \\[2mm]
&= 2\pi (2 - 0.5) \\[2mm]
&= 3\pi \ \text{units}^{3}
\end{aligned}
$$

◊

In the next example we will try and find the volume enclosed by rotating a curve about the $y$ axis. This means we will need to be much more careful about the heights of the cylinders.

**Example 3.4**   Find the volume enclosed by rotating the area bounded by $f(x) = x^{2/3}$, $x = 1$, and $y = 3^{2/3}$ about the $y$-axis from $x = 1$ to $x = 3$.

**Solution:**

Start with the traditional sketch.

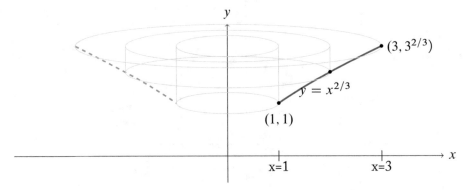

This example is one where the height of the cylinder is not just $f(x)$, so we need to find an expression for the height of the cylinder. This expression goes from $f(x)$ (at the bottom) to $y = 3^{2/3}$ (at the top). So, we get a formula of

$$h = 3^{2/3} - f(x) = 3^{2/3} - x^{2/3}$$

for the height of the cylinder. The area of a cylinder is thus $2\pi x \cdot \left(3^{2/3} - x^{2/3}\right)$ which makes the volume:

$$
\begin{aligned}
\text{Volume} &= 2\pi \int_1^3 x \cdot \left(3^{2/3} - x^{2/3}\right) dx \\
&= 2\pi \int_1^3 \left(x \cdot 3^{2/3} - x \cdot x^{2/3}\right) dx \\
&= 2\pi \int_1^3 \left(3^{2/3}x - x^{5/3}\right) dx \\
&= 2\pi \left(\frac{3^{2/3}}{2}x^2 - \frac{3}{8}x^{8/3}\right)\Bigg|_1^3 \\
&\cong 10.52 \text{ units}^3
\end{aligned}
$$

◊

Example 3.4 demonstrates that the formula in Knowledge Box 3.2 only covers one type of rotation about the $y$ axis. In general, it is necessary to remember that you are adding up cylindrical shells and to carefully figure out the height and radius, plugging into $A = 2\pi r h$ to get the formula to integrate. Care is also needed in figuring out the limits of integration. Next, we provide an example of a problem where we use disks to rotate about the $y$ axis.

In order to do this we need to use inverse functions. These were defined in *Fast Start Differential Calculus*. Examples of inverse functions include the following.

- For $x \geq 0$, when $f(x) = x^2$ we have $f^{-1}(x) = \sqrt{x}$.

- If $g(x) = e^x$, we have $g^{-1}(x) = \ln(x)$.

- If $h(x) = \tan(x)$, we have $h^{-1}(x) = \tan^{-1}(x)$.

The notation for "inverse" and "negative first power" are identical and can only be told apart by examining context. Be careful!

**Example 3.5    Rotation about the $y$-axis with disks:** Find the volume of rotation when the area bounded by $y = x^2$, the $y$-axis, and $y = 4$ is rotated about the $y$ axis.

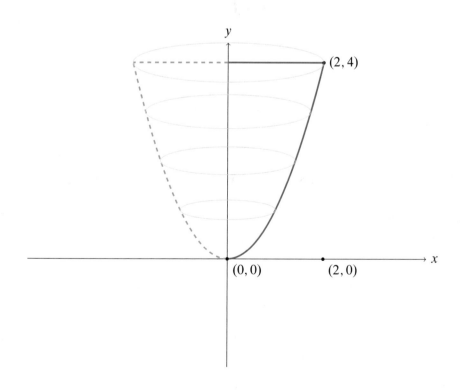

**Solution:**

For this problem, the radius of the disks is going from the $y$-axis to the curve $y = x^2$. We need to solve this equation for $x$ which is not too difficult:

$$x = g(y) = \sqrt{y}$$

We ignore the negative square root because we obviously need a positive radius. This means that we have a disk radius of $x = g(y)$.

The volume formula becomes

$$V = \pi \int_{y_0}^{y_1} g(y)^2 \cdot dy$$

The function $x = g(y)$ is the *inverse function* of $y = f(x)$. We can now do the calculations for volume:

$$\text{Volume} = \pi \int_0^4 (\sqrt{y})^2 \cdot dy$$

$$= \pi \int_0^4 y \cdot dy$$

$$= \frac{\pi}{2} y^2 \Big|_0^4$$

$$= \frac{\pi}{2}(16 - 0)$$

$$= 8\pi \ \text{units}^3$$

To see that is it possible, let's set this up (but not calculate the integral) with cylinders.

The height of a cylinder is from $y = x^2$ to $y = 4$ meaning that $h = 4 - x^2$. So, the volume integral is

$$V = 2\pi \int_{x=0}^{x=2} x(4 - x^2) \cdot dx$$

Not a terribly hard integral, but one that is harder than the disk integral.

$\Diamond$

**Knowledge Box 3.3**

## Disks or Cylinders?

*Both methods for finding volume of rotation involve adding up areas with integration to find a volume. The method of disks adds up circular disks and the method of cylinders adds up, well, cylinders. How do you tell which method to use?*

**You use whichever method you can set up and, if you can set up both, you use the one that yields the easier integral.**

The next example is another one that lets us practice with the method of cylinders. It is an example where the calculations to find the radius values needed to use the method of disks is too hard.

**Example 3.6**    Compute the volume obtained by rotating the area bounded by the curve

$$y = 2x^2 - x^3$$

and the $x$-axis around the $y$-axis.

**Solution:**

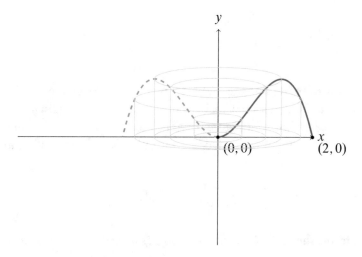

Let's do the calculations. The points where $y = 0$ are $x = 0$ and $x = 2$, giving us the limits of integration. In this case the radius of the cylinders is just $x$, and the height is just $y$. So, applying the formula from Knowledge Box 3.2 we get:

$$\text{Volume} = 2\pi \int_0^2 x(2x^2 - x^3) \cdot dx$$

$$= 2\pi \int_0^2 (2x^3 - x^4) \cdot dx$$

$$= 2\pi \left( \frac{1}{2}x^4 - \frac{1}{5}x^5 \Big|_0^2 \right)$$

$$= 2\pi \left( 8 - \frac{32}{5} - 0 + 0 \right)$$

$$= 2\pi \cdot \frac{8}{5}$$

$$= \frac{16\pi}{5} \ \text{units}^3$$

If we wanted to use the method of disks we would need to solve $y = 2x^2 - x^3$ for $x$ to get the inverse function – a challenging piece of algebra.

◊

Suppose that we want to rotate the area between two curves about the $x$-axis. Then we get a large disk for the outer curve and a small disk for the inner curve – meaning that we get *washers*. A rendering of a washer is shown in Figure 3.4. The area of a washer with outer radius $r_1$ and inner radius $r_2$ is the difference of the area of the overall disk and the missing inner disk:

$$A = \pi r_1^2 - \pi r_2^2 = \pi(r_1^2 - r_2^2)$$

This area formula forms the basis of the integration performed with the **method of washers**.

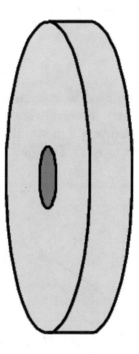

Figure 3.4: A washer or disk with a hole in the center. When we want to rotate the difference of two curves about the $x$-axis, this shape replaces the disks used when only one curve is rotated.

### Knowledge Box 3.4

**Formula for finding volume with the method of washers**

*If $y = f_1(x)$ and $y = f_2(x)$ are functions for which $f_1(x) \geq f_2(x)$ on the interval $[a, b]$, then the volume obtained by rotating the area between the curves about the $x$-axis is:*

$$V = \pi \int_a^b \left( f_1(x)^2 - f_2(x)^2 \right) \cdot dx.$$

*(Notice that we took the constant $\pi$ out in front of the integral sign.)*

**Example 3.7**   Find the volume resulting from rotating the area between $f_1(x) = x^2$ and $f_2(x) = \sqrt{x}$ about the $x$-axis.

**Solution:**

To apply the method of washers, we need to know where the curves intersect and which one is on the outside edge of the washers. It is easy to see that they intersect at $(0, 0)$ and $(1, 1)$. So the limits of integration will be from $x = 0$ to $x = 1$. On the range $0 \leq x \leq 1$, it's easy to see that $x^2 \leq \sqrt{x}$. So, the outer curve is $f_2(x)$. This gives us enough information to set up the integral.

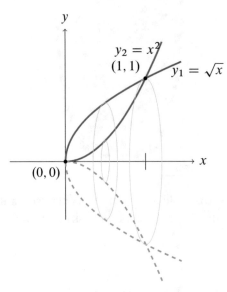

$$\text{Volume} = \pi \int_0^1 \left( f_1(x)^2 - f_2(x)^2 \right) \cdot dx$$

$$= \pi \int_0^1 \left( x - x^4 \right) dx$$

$$= \pi \left. \left( \frac{1}{2}x^2 - \frac{1}{5}x^5 \right) \right|_0^1$$

$$= \frac{3\pi}{10} \; \text{units}^3$$

◇

There are a large number of different ways to rotate objects about the $x$- and $y$-axis. While formulas are presented in this section, it is a really good idea to identify the radius and height of the disks, cylindrical shells, or washers you are integrating to create a volume of rotation. Without that understanding, it is easy to pick the wrong formula. For Example 3.4 *none* of the formulas apply!

The formulas given in this section demonstrate how to add up infinitely thin disks, cylinders, and washers with an integral to compute a volume. These are relatively intuitive shapes with which we should already be familiar. Calculus is not constrained by familiarity: an integral can add up any sort of shapes. If you want a challenge, try to find the formula for the volume of a pyramid with a square base. The formula should be based on the side length of the square base and the height of the pyramid.

## PROBLEMS

**Problem 3.8**    Find the volume of rotation about the $x$-axis for each of the functions below from $x = 0$ to $x = 1$.

1. $f(x) = 3x^5$

2. $g(x) = 0.5x^6$

3. $h(x) = 5x^7$

4. $r(x) = 12x^5$

5. $s(x) = 6x^{11}$

6. $q(x) = e^{-x}$

7. $a(x) = \cos(\pi x)$

8. $b(x) = (e^x + e^{-x})$

**Problem 3.9**    If we rotate $y = \sin(x)$ about the $x$-axis, the result is a string of beads. Find the volume of one bead.

**Problem 3.10**    Repeat Problem 3.9 for $f(x) = \sin(4x)$.

**Problem 3.11**    Recalculate Example 3.5 using the method of cylinders.

**Problem 3.12**    Verify the volume formula

$$V = \frac{4}{3}\pi r^3$$

for a sphere using the method of disks about the $x$-axis.

**Problem 3.13**    Find the value $a$ so that rotating the curve $y = \ln(x)$ about the $x$-axis from $x = 0$ to $x = a$ yields a volume of 4 units$^3$.

**Problem 3.14**   Find the volume of rotation about the $y$ axis of the area below each of the functions listed from $x = 0$ to $x = 1$.

1. $f(x) = 6x^9$

2. $g(x) = 16x^8$

3. $h(x) = 3x^{12}$

4. $r(x) = 5x^8$

5. $s(x) = 11x^8$

6. $q(x) = e^x$

7. $a(x) = \cos(\pi x)$

8. $b(x) = \frac{x}{x^2+1}$

**Problem 3.15**   We know the area under $y = 1/x$ on the interval $[1, \infty)$ is infinite. Find the volume of rotation of $y = 1/x$ on this interval.

**Problem 3.16**   For each of the following functions and starting and ending $x$ values, find the volume of rotation of the function about the $x$-axis.

1. $f(x) = x^{-2/5}$ between $x = 3$ and $x = 7$

2. $f(x) = x^{-2/5}$ between $x = 3$ and $x = 5$

3. $f(x) = x^{2/7}$ between $x = 4$ and $x = 7$

4. $f(x) = x^{3/7}$ between $x = 5$ and $x = 7$

5. $f(x) = x^{-3}$ between $x = 5$ and $x = 9$

6. $f(x) = x^{-2}$ between $x = 8$ and $x = 9$

**Problem 3.17**   For each of the following pairs of functions, compute the volume obtained by rotating the area between the functions about the $x$-axis.

1. $f(x) = 3x^3$ and $g(x) = 9x^2$

2. $f(x) = x^3$ and $g(x) = 4x^2$

3. $f(x) = 3x^5$ and $g(x) = 48x$

4. $f(x) = 3x^4$ and $g(x) = 3x^3$

5. $f(x) = 2x^3$ and $g(x) = 6x^2$

6. $f(x) = x^3$ and $g(x) = x^2$

**Problem 3.18**    For each of the following pairs of functions, compute the volume obtained by rotating the area between the functions about the $y$-axis.

1. $f(x) = 7x^3$ and $g(x) = 28x^2$

2. $f(x) = 4x^6$ and $g(x) = 64x^2$

3. $f(x) = 2x^4$ and $g(x) = 18x^2$

4. $f(x) = 3x^6$ and $g(x) = 3x^2$

5. $f(x) = 7x^5$ and $g(x) = 21x^4$

6. $f(x) = 2x^3$ and $g(x) = 8x$

**Problem 3.19**    Using the method of disks, rotating about the $x$-axis, verify the formula for the volume of a cone of radius $R$ and height $H$:

$$V = \frac{1}{3}\pi \cdot R^2 H$$

## 3.2    ARC LENGTH AND SURFACE AREA

In this section we will learn to compute the length of curves and, having done that, to find the surface area of figures of rotation.

A piece of a curve is called an **arc**. The key to finding the length of an arc is the **differential of arc length**.

In the past we have had quantities like $dx$ and $dy$ that measure infinitesimal changes in the directions of the variables $x$ and $y$.

The differential of arc length is different – it does not point in a consistent direction, rather it points along a curve and so, by integrating it, we can find the length of a curve.

Examine Figure 3.5. The relationship between the change in $x$ and $y$ and the change in the length of the curve is Pythagorean, based on a right triangle.

If we take this relationship to the infinitesimal scale, we obtain a formula for the differential of arc length.

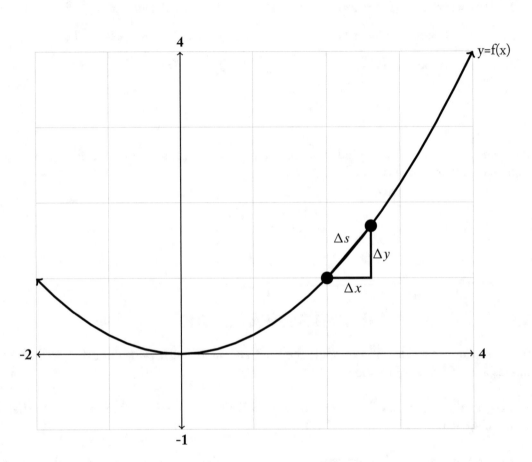

Figure 3.5: The triangle with sides $\Delta x$, $\Delta y$, and $\Delta s$ shows how the change in the length of the curve is related to the changes in distance in the $x$ and $y$ directions.

**Knowledge Box 3.5**

### The differential of arc length and arc length

*If $y = f(x)$ is a continuous curve, then the rate at which the length of the graph of $f(x)$ changes is called the* **differential of arc length,** *denoted by $ds$. The value of $ds$ is:*

$$ds^2 = dy^2 + dx^2$$

$$ds = \sqrt{dy^2 + dx^2}$$

$$= \sqrt{\left(\frac{dy^2}{dx^2} + 1\right) \cdot dx^2}$$

$$= \sqrt{(y')^2 + 1} \cdot dx.$$

*The length $S$ of a curve (**arclength** of the curve) from $x = a$ to $x = b$ is:*

$$S = \int_a^b ds.$$

**Example 3.20**   Find the length of $y = 3x^{2/3}$ from $x = 1$ to $x = 8$.

**Solution:**

The first step in an arc length problem is to compute $ds$.

$$y = 3x^{2/3}$$

$$y' = 2x^{-1/3}$$

$$ds = \sqrt{4x^{-2/3} + 1} \cdot dx$$

This means that the desired length is

$$S = \int_1^8 ds$$

$$= \int_1^8 \sqrt{\frac{4}{x^{2/3}} + 1} \cdot dx$$

$$= \int_1^8 \sqrt{\frac{4 + x^{2/3}}{x^{2/3}}} \cdot dx$$

$$= \int_1^8 \sqrt{4 + x^{2/3}} \cdot \frac{dx}{x^{1/3}}$$

Let $u = 4 + x^{2/3}$, then $du = \frac{2}{3}x^{-1/3} \cdot dx = \frac{2}{3}\frac{dx}{x^{1/3}}$

So, $\frac{3}{2}du = \frac{dx}{x^{1/3}}$.

Applying the substitution to the limits we see that the integral goes from $u = 5$ to $u = 8$. Transforming everything to $u$-space, the arc length is:

$$S = \int_5^8 \sqrt{u} \cdot \frac{3}{2}du$$

$$= \frac{3}{2}\int_5^8 u^{1/2}du$$

$$= \frac{3}{2}\left(\frac{2}{3}u^{3/2}\right)\Big|_5^8$$

$$= 8^{3/2} - 5^{3/2}$$

$$\cong 11.45 \text{ units}^2$$

$$\Diamond$$

Alert students will have noticed that the function chosen to demonstrate arc length is not one of our usual go-to functions for demonstration. This is because the formula for $ds$ yields some very difficult integrals. The next example is one such, but yields a formula we already know how to integrate.

**Example 3.21**   Find the arc length of $y = x^2$ from $x = 0$ to $x = 2$.

**Solution:**

Since $y' = 2x$, it is easy to find that $ds = \sqrt{4x^2 + 1} \cdot dx$, meaning our integral is:

$$S = \int_0^2 \sqrt{4x^2 + 1} \cdot dx$$

This is a trig-substitution integral. The triangle for this integral is

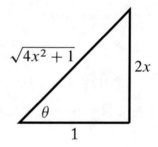

which means our substitutions are:

$$\sqrt{4x^2 + 1} = \sec(\theta)$$

$$2x = \tan(\theta)$$

$$x = (1/2)\tan(\theta)$$

$$dx = (1/2)\sec^2(\theta) \cdot d\theta$$

So we get:

$$\int_0^2 \sqrt{4x^2 + 1} \cdot dx = \int_?^? \sec(\theta) \cdot (1/2) \sec^2(\theta) \cdot d\theta$$

$$= \frac{1}{2} \int_?^? \sec^3(\theta) \cdot d\theta \qquad\qquad \text{(known integral)}$$

$$= \frac{1}{4} \left( \sec(\theta) \tan(\theta) + \ln|\sec(\theta) + \tan(\theta)| \right) \Big|_?^?$$

$$= \frac{1}{4} \left( \sqrt{4x^2 + 1} \cdot 2x + \ln|\sqrt{4x^2 + 1} + 2x| \right) \Big|_0^2$$

Now that we have performed the integral and transformed it back into $x$-space we can substitute in the limits and get the arc length.

$$S = \frac{1}{4} \left( \sqrt{17} \cdot 4 + \ln|\sqrt{17} + 4| - \sqrt{1} \cdot 0 + \ln|\sqrt{1} + 0| \right)$$
$$= \frac{1}{4} \left( 4\sqrt{17} + \ln(\sqrt{17} + 4) \right) \cong 4.65 \text{ units}$$

Arc-lengths integrals are often challenging. Let's do one more example with a function chosen to keep the difficulty from getting out of hand. This example looks for a general formula for the arc length of a function.

**Example 3.22**   Find the length of $y = x^{3/2}$ from $x = 0$ to $x = a$.

**Solution:**

Since $y' = \frac{3}{2}x^{1/2}$ we see that

$$ds = \sqrt{\frac{9}{4}x + 1} \cdot dx = \frac{1}{2}\sqrt{9x + 4} \cdot dx$$

This means that the desired arc length is:

$$S = \int_0^a \frac{1}{2} \sqrt{9x + 4} \cdot dx$$

$$= \frac{1}{2} \int_0^a \sqrt{9x + 4} \cdot dx$$

$$\text{Let } u = 9x + 4 \text{ so that } \frac{1}{9} du = dx$$

$$= \frac{1}{2} \int_?^? \sqrt{u} \cdot \frac{1}{9} du$$

$$= \frac{1}{18} \int_?^? u^{1/2} \cdot du$$

$$= \frac{1}{18} \left( \frac{2}{3} u^{3/2} \right) \Big|_?^?  \qquad\qquad \text{Need to substitute back to } x$$

$$= \frac{1}{27} (9x + 4)^{3/2} \Big|_0^a$$

$$= \frac{(9a + 4)^{3/2} - 8}{27}$$

which is the desired arc length formula.

We have already computed the volume of a solid that is enclosed by the graph of a function rotated about the $x$-axis. The solids defined in this fashion also have a surface area, the slices of which are circles.

Figure 3.6 shows examples of the circles that appear in such a rotation.

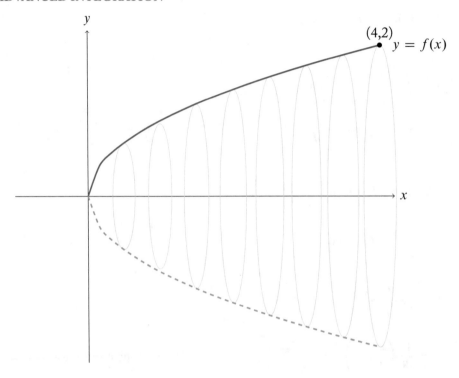

Figure 3.6: Shown are some circles that are the "slices" of the surface area obtained by rotating $f(x)$ about the $x$-axis.

Using Figure 3.6 as inspiration, we can compute surface area by integrating the circumference of circles with radius $f(x)$. Since these circle follow the arc of $f(x)$, the correct type of change is the differential of arc length, $ds$. This leads to the formula in Knowledge Box 3.6.

### Knowledge Box 3.6

#### Surface area of rotation

*If $y = f(x)$ is a continuous curve, then the surface area $A$ obtained by rotating $f(x)$ around the $x$-axis from $x = a$ to $x = b$ is:*

$$A = 2\pi \int_a^b f(x) \cdot ds.$$

**Example 3.23**   Find the surface area of rotation of $y = \sqrt{x}$ about the $x$-axis from $x = 0$ to $x = 4$.

**Solution:**

Since $y = x^{1/2}$, $y' = \dfrac{1}{2}x^{-1/2} = \dfrac{1}{2\sqrt{x}}$. So: $ds = \sqrt{\dfrac{1}{4x} + 1} \cdot dx$ Using the formula from Knowledge Box 3.6 we obtain the surface area integral,

$$A = 2\pi \int_0^4 \sqrt{x} \cdot \sqrt{\frac{1}{4x} + 1} \cdot dx$$

$$= 2\pi \int_0^4 \sqrt{x \cdot \left(\frac{1}{4x} + 1\right)} \dot{d}x$$

$$= 2\pi \int_0^4 \sqrt{\frac{1 + 4x}{4}} \cdot dx$$

$$= 2\pi \int_0^4 \frac{1}{2}\sqrt{1 + 4x} \cdot dx$$

$$= \pi \int_0^4 \sqrt{1 + 4x} \cdot dx$$

$$\text{Let } u = 4x + 1, \ \frac{1}{4}du = dx$$

$$= \pi \int_1^{17} u^{1/2} \cdot \frac{1}{4}du$$

$$= \frac{\pi}{4} \int_1^{17} u^{1/2} \cdot du$$

$$= \frac{\pi}{4}\frac{2}{3} u^{3/2}\Big|_1^{17}$$

$$= \frac{\pi}{6}\left(17^{3/2} - 1\right) \cong 36.18 \text{ units}^2$$

$\diamond$

**Example 3.24**   Find the surface area of rotation for $y = e^x$ from $x = 0$ to $x = 2$.

**Solution:**

Noting that $y' = e^x$, we see that

$$ds = \sqrt{e^{2x} + 1} \cdot dx.$$

This means the surface area is:

$$A = 2\pi \int_0^2 e^x \sqrt{e^{2x} + 1} \cdot dx$$

$$= 2\pi \int_0^2 \sqrt{e^{2x} + 1} \, (e^x \cdot dx)$$

Let $u = e^x$, then $du = e^x \cdot dx$

$$= 2\pi \int_?^? \sqrt{u^2 + 1} \cdot du \qquad\qquad \text{(known integral)}$$

$$= 2\pi \left( u \sqrt{u^2 + 1} + \ln|u + \sqrt{u^2 + 1}| \right)\Big|_?^?$$

$$= 2\pi \left( e^x \sqrt{e^{2x} + 1} + \ln|e^x + \sqrt{e^{2x} + 1}| \right)\Big|_0^2$$

$$= 2\pi \left( e^2 \sqrt{e^4 + 1} + \ln(e^2 + \sqrt{e^4 + 1}) - \sqrt{2} - \ln|1 + \sqrt{2}| \right) \text{ units}^2$$

$$\Diamond$$

The types of integrals that arise from arc length and rotational surface area problems are often quite challenging. This justifies the large number of integration techniques we learned when we studied methods of integration, several of which came up in this section.

If you study multivariate calculus more deeply, the differential of arc length will appear again for tasks like computing the work done moving a particle along a path through a field. This section is a bare introduction to the power and applications of the differential of arc length.

# PROBLEMS

**Problem 3.25**    For each of the following functions, compute $ds$, the differential of arc length.

1. $f(x) = x^3$

2. $g(x) = \sin(x)$

3. $h(x) = \tan^{-1}(x)$

4. $r(x) = \dfrac{x}{x+1}$

5. $s(x) = e^{-x}$

6. $q(x) = x^{3/4}$

7. $a(x) = \dfrac{1}{3}$

8. $b(x) = 2^x$

**Problem 3.26**    For each of the following functions, compute the arc length of the graph of the function on the given interval.

1. $f(x) = 9x^{2/3}$; $[0, 1]$

2. $g(x) = 2x + 1$; $[0, 4]$

3. $h(x) = 2x^{3/2}$; $[2, 5]$

4. $r(x) = x^2 + 4x + 4$; $[0, 6]$

5. $s(x) = \sqrt{(x-2)^3}$; $[4, 5]$

6. $q(x) = (x+1)^{2/3}$; $[-1, 1]$

7. $a(x) = 4x^{3/2}$; $[2, 4]$

8. $b(x) = x^2 + x + 1/4$; $[0, 8]$

**Problem 3.27**    For each of the following functions, compute the surface area of rotation of the function for the given interval.

1. $f(x) = x$; $[0, 3]$

2. $g(x) = \sqrt{x}$; $[4, 9]$

3. $h(x) = e^x$; $[0, 1]$

4. $r(x) = \sin(x)$; $[0, \pi]$

5. $s(x) = \cos(3x)$; $[0, \pi/4]$

6. $q(x) = \sin(x)\cos(x)$; $[0, \pi/2]$

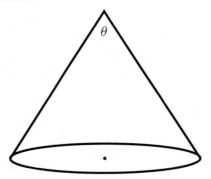

**Problem 3.28**   Using the techniques for surface area of revolution, find the formula for the surface area of a cone with apex angle $\theta$, as shown above. Don't forget the area of the bottom.

**Problem 3.29**   If
$$y = x^{2/3},$$
find the formula for the arc length of the graph of this function on the interval $[a, b]$.

**Problem 3.30**   Based on the material in this section, if
$$(f(t), g(t))$$
is a parametric curve, what would the differential of arc length, $ds$, be?

**Problem 3.31**   Derive the polar differential of arc length.

**Problem 3.32**   Derive the parametric differential of arc length for
$$(x(t), y(t))$$

**Problem 3.33**   Find but do not evaluate the integral for computing the arc length of
$$y = \sin(x).$$
Discuss: what techniques might work for this integral.

**Problem 3.34** Which is harder, finding the arc-length of

$$y = \sqrt{x}$$

or its surface area of rotation? Why?

**Problem 3.35** We already know the area under $y = 1/x$ on the interval $[1, \infty)$ is infinite but that the enclosed volume of rotation is finite. Using comparison and cleverness, demonstrate the surface area of this shape is infinite.

## 3.3 MULTIPLE INTEGRALS

In our earlier study of integration, we learned to use integrals to find the area under a curve. The analogous task in three dimensions is to find the volume under a surface over some domain in the $x$-$y$ plane. As with partial derivatives we will find the idea of a *currently active* variable useful.

When we were integrating a single-variable function to obtain an area, we integrated over an interval on the $x$-axis. When we are finding volumes under surfaces, we will integrate over *regions* or subsets of the plane. An example of a relatively simple region in the plane is shown in Figure 3.7.

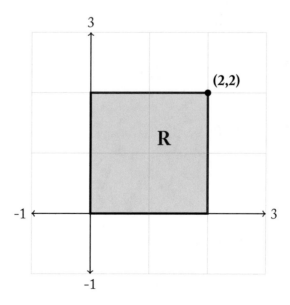

Figure 3.7: A rectangular region $R$, $0 \leq x, y \leq 2$ in the plane.

So far in this chapter we have found volumes by using integrals to add up slices. To perform volume integrals we will have to integrate slices that are, themselves, the result of integrating. This means we will use **multiple or iterated integrals**. Fortunately, these are done one at a time. We use differentials, like $dy$ and $dx$, to cue which integral is performed next. Like forks at a fancy dinner party, the next integral is the one for the differential farthest to the left.

**Example 3.36**   Find the volume under

$$f(x, y) = x^2 + y^2$$

that is above $R$, $0 \le x, y \le 2$, as shown in Figure 3.7.

**Solution:**

The $y$ and $x$ variables are both in intervals of the form $[0, 2]$, so the limits of integration are 0 and 2 for both the integrals we must perform. The volume under the surface is:

$$V = \int_0^2 \int_0^2 (x^2 + y^2) \cdot dx \, dy \qquad \text{The variable } x \text{ is active.}$$

$$= \int_0^2 \left( \frac{1}{3}x^3 + xy^2 \right) \Big|_0^2 \cdot dy$$

$$= \int_0^2 \left( \frac{8}{3} + 2y^2 - 0 - 0 \right) \cdot dy$$

$$= \int_0^2 \left( 2y^2 + \frac{8}{3} \right) \cdot dy \qquad \text{Now } y \text{ is active.}$$

$$= \frac{2}{3}y^3 + \frac{8}{3}y \Big|_0^2$$

$$= \left( \frac{16}{3} + \frac{16}{3} - 0 - 0 \right) = \frac{32}{3} \text{ units}^3$$

◇

Example 3.36 contained two integrals. During the first, $x$ was the active variable, and $y$ acted like a constant; when the limits of integration were substituted in they were substituted in for $x$

*not* $y$. The second integral took place in an environment where $x$ was gone and $y$ was the active variable.

**Knowledge Box 3.7**

**Integration with respect to a variable**

*When we are computing an integral*

$$V = \int_a^b \int_c^d f(x, y) \cdot dx \ dy,$$

*the first integral treats $x$ as a variable and is said to be an integral* **with respect to $x$;** *the second integral treats $y$ as a variable and is said to be an integral* **with respect to $y$.**

When we are integrating with respect to one variable, other variables act as constants and so can pass through integral signs. This permits us to simplify some integrals.

**Example 3.37**   Use the constant status of variables to compute

$$\int_0^1 \int_0^1 (x^2 y^2) \cdot dx \ dy$$

in a simple way.

**Solution:**

$$\int_0^1 \int_0^1 (x^2 y^2) \cdot dx \ dy = \left( \int_0^1 y^2 \cdot dy \int_0^1 x^2 \cdot dx \right) \qquad \text{Pass } y \text{ through the } x \text{ integral}$$

$$= \left( \int_0^1 y^2 \cdot dy \right) \cdot \left( \int_0^1 x^2 \cdot dx \right)$$

$$= \left( \int_0^1 x^2 \cdot dx \right)^2 \qquad \text{Since the integrals are equal}$$

$$= \left( \frac{1}{3} x^3 \Big|_0^1 \right)^2$$

$$= \left( \frac{1}{3} - 0 \right)^2 = \frac{1}{9} \text{ units}^3$$

◊

This sort of integral – that can be split up into two different integrals – is called a *decomposable integral.*

**Knowledge Box 3.8**

**Decomposable Integrals**

*The multiple integral of the product of a function of one variable by a function of the other variable can be factored into two single-variable integrals.*

$$\int \int f(x)g(y) \cdot dx \, dy = \left( \int f(x) \, dx \right) \left( \int g(y) \, dy \right).$$

**Example 3.38**   Use the decomposition of integrals to perform the following:

$$\int_0^{\pi/2} \int_0^2 (x \cdot \cos(y)) \cdot dx \, dy$$

**Solution:**

$$\int_0^{\pi/2} \int_0^2 (x \cdot \cos(y)) \cdot dx \, dy = \left( \int_0^2 x \, dx \right) \times \left( \int_0^{\pi/2} \cos(y) \, dy \right)$$

$$= \left( \frac{x^2}{2} \Big|_0^2 \right) \times \left( \sin(y) \Big|_0^{\pi/2} \right)$$

$$= (4/2 - 0/2) \times (\sin(\pi/2) - \sin(0)) = 2 \cdot 1 = 2$$

◊

Volume integration becomes more difficult when the region $R$ is not a rectangle. Over a rectangular region, the limits of integration are constants. If a region is not rectangular, then the curves that describe the boundaries of the region become involved in the limits of integration.

**Example 3.39** Integrate the function $f(x, y) = 2x - y + 4$ over the region $R$ given in Figure 3.8.

**Solution:**

The function is simple, but the limits of integration are tricky. In this case, $0 \le x \le 2$ and, *for a given value of $x$*, $0 \le y \le x$. This is because the upper edge of the region of integration is the line $y = x$.

Now, we can set up the integral, choosing the order of integration to agree with the limits.

$$= \int_0^2 \int_0^x (2x - y + 4)\, dy\, dx$$

$$= \int_0^2 \left( 2xy - \frac{1}{2}y^2 + 4y \right) \Big|_0^x dx$$

$$= \int_0^2 \left( 2x^2 - \frac{1}{2}x^2 + 4x - 0 - 0 - 0 \right) \cdot dx$$

$$= \int_0^2 \left( \frac{3}{2}x^2 + 4x \right) dx$$

$$= \frac{1}{2}x^3 + 2x^2 \Big|_0^2$$

$$= \frac{1}{2}8 + 8 - 0 - 0$$

$$= 12 \text{ units}^3$$

◊

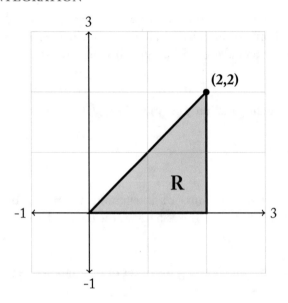

Figure 3.8: **A** non-rectangular region bounded by the $x$-axis, the line $x = 2$, and the line $y = x$.

**Example 3.40**   Find the volume underneath

$$f(x, y) = 4 - x^2$$

over the region bounded by

$$y = x$$

$$y = 2x$$

and

$$x = 2$$

**Solution:**

Start by drawing the region of integration.

This region has $x$ bounds $0 \leq x \leq 2$.

For a given value of $x$ the region goes from the line $y = x$ to the line $y = 2x$.

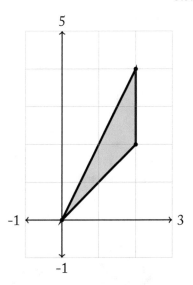

This gives us sufficient information to set up the integral.

$$V = \int_0^2 \int_x^{2x} \left(4 - x^2\right) dy\, dx$$

$$= \int_0^2 \left(4y - x^2 y\right) \Big|_x^{2x} \cdot dx$$

$$= \int_0^2 \left(8x - 2x^3 - 4x + x^3\right) dx$$

$$= \int_0^2 \left(4x - x^3\right) dx$$

$$= 2x^2 - \frac{1}{4}x^4 \Big|_0^2$$

$$= 8 - \frac{1}{4} \cdot 16 - 0 + 0$$

$$= 4 \text{ units}^3$$

◊

The regions we have used thus far have been based on functions that are easy to work with using Cartesian coordinates.

To deal with other sorts of regions, we need to first develop a broader point of view.

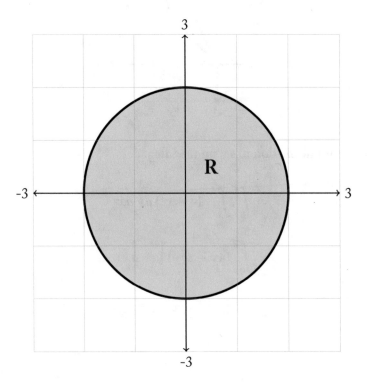

Figure 3.9: Another sort of region – a disk of radius 2 centered at the origin.

**Knowledge Box 3.9**

**The differential of area**

*The* **change of area** *is*

$$dA = dx \, dy = dy \, dx.$$

*This permits us to change our integral notation to the following for the integral of $f(x, y)$ over a region $R$*

$$V = \int \int_R f(x, y) \, dA.$$

*In polar coordinates:*

$$dA = r \, dr \, d\theta.$$

**Example 3.41**    Find the area enclosed below the curve $f(x, y) = x^2 + y^2$ over a disk of radius 2 centered at the origin.

**Solution:**

This volume is below the function $f(x, y)$ for points no farther from the origin than 2. This means the region $R$ is

$$x^2 + y^2 \leq 4.$$

In polar coordinates, this region is those points $(r, \theta)$ for which $0 \leq r \leq 2$ and $0 \leq \theta \leq 2\pi$. Since the polar/rectangular conversion equations tell us $r^2 = x^2 + y^2$, $f(r, \theta) = r^2$.

Using polar coordinates the integral is:

$$V = \int \int_R f(r, \theta) \, dA$$

$$= \int_0^{2\pi} \int_0^2 r^2 \cdot r \cdot dr \, d\theta$$

$$= \int_0^{2\pi} \int_0^2 r^3 \cdot dr \, d\theta$$

$$= \int_0^{2\pi} \frac{1}{4} r^4 \Big|_0^2 \, d\theta$$

$$= \int_0^{2\pi} (16/4 - 0) \, d\theta$$

$$= \int_0^{2\pi} 4 \, d\theta$$

$$= 4\theta \Big|_0^{2\pi}$$

$$= 8\pi - 0 = 8\pi \text{ units}^3$$

◇

The next example is a very important one for the theory of statistics. As you know if you have studied statistics, the normal distribution has a probability distribution function of:

$$\frac{1}{\sqrt{2\pi}} e^{-x^2/2}$$

The area under the curve of a probability distribution function must be equal to one. Thus, if you have a function with an area greater than one, you must multiply it by a normalizing constant equal to one over the area.

The following example shows where the normalizing constant $\frac{1}{\sqrt{2\pi}}$ in the normal distribution probability distribution function comes from.

The integral relies on a trick: squaring the integral and then shifting the squared integral to polar coordinates. This changes an impossible integral into one that can be done without difficulty by $u$-substitution. Sadly, this only permits the evaluation of the integral on the interval $[-\infty, \infty]$; the coordinate change is intractable *except* on the full interval where the function exists.

**Example 3.42** Find $A = \int_{-\infty}^{\infty} e^{-x^2/2} \cdot dx$.

**Solution:**

The solution works if you compute the square of the integral.

$$A^2 = \left( \int_{-\infty}^{\infty} e^{-x^2/2} \cdot dx \right)^2$$

$$= \left( \int_{-\infty}^{\infty} e^{-x^2/2} \cdot dx \right) \left( \int_{-\infty}^{\infty} e^{-x^2/2} \cdot dx \right)$$

$$= \left( \int_{-\infty}^{\infty} e^{-x^2/2} \cdot dx \right) \left( \int_{-\infty}^{\infty} e^{-y^2/2} \cdot dy \right) \qquad \text{Rename}$$

$$= \int_{-\infty}^{\infty} \int_{-\infty}^{\infty} e^{-x^2/2} e^{-y^2/2} \cdot dy \, dx$$

$$= \int_{-\infty}^{\infty} \int_{-\infty}^{\infty} e^{-1/2(x^2+y^2)} \cdot dA$$

Change to polar coordinates

$$= \int \int_R e^{-1/2(r^2)} \cdot dA$$

$$= \int \int_R e^{-1/2(r^2)} \cdot r \cdot dr \, d\theta$$

The polar region in question is $0 \le r < \infty$ and $0 \le \theta < 2\pi$. Rebuild the integral with these limits and we get:

$$A^2 = \int_0^{2\pi} \int_0^{\infty} r \cdot e^{-r^2/2} \cdot dr \, d\theta$$

$$= \left( \int_0^{2\pi} d\theta \right) \cdot \left( \int_0^{\infty} r \cdot e^{-r^2/2} \cdot dr \right)$$

$$= \theta \Big|_0^{2\pi} \cdot \left( \int_0^{\infty} r \cdot e^{-r^2/2} \cdot dr \right)$$

$$= 2\pi \int_0^\infty r \cdot e^{-r^2/2} \cdot dr$$

$$= 2\pi \lim_{a \to \infty} \int_0^a r \cdot e^{-r^2/2} \cdot dr$$

Let $u = -r^2/2$, then $-du = r \cdot dr$

$$= 2\pi \lim_{a \to \infty} \int_?^? e^u \cdot -du$$

$$= -2\pi \lim_{a \to \infty} \int_?^? e^u \cdot du$$

$$= -2\pi \lim_{a \to \infty} \left. e^{-r^2/2} \right|_0^a$$

$$= -2\pi \lim_{a \to \infty} \left( e^{-a^2/2} - 1 \right)$$

$$= -2\pi (0 - 1) = 2\pi$$

If $A^2 = 2\pi$ then $A = \sqrt{2\pi}$, which is the correct normalizing constant.

### 3.3.1   MASS AND CENTER OF MASS

The center of mass for an object is the average position of all the mass in an object. This section demonstrates techniques for computing the center of mass of flat plates with a density function $\rho(x, y)$. Density is the rate at which mass changes as you move through an object, which, in turn, means that the mass of an object is the integral of its density.

**Knowledge Box 3.10**

## Mass of a plate

*Suppose that a flat plate occupies a region R with a density function $\rho(x, y)$ defined on R. Then the mass of the plate is*

$$M = \int \int_R \rho(x, y) \cdot dA.$$

Remember that the function $\rho(x, y)$ is usually constant, or close enough to constant that we assume it to be constant, when we have a mass made of a relatively uniform material. The fairly high variation in the mass functions in the examples and homework problems is intended to give your integration skills a workout – not as a representation of situations encountered in physical reality.

**Example 3.43**   If a plate fills the triangular region $R$ from Figure 3.8 with a density function $\rho(x, y) = x + 1$ grams/unit$^2$, find the mass of the plate.

**Solution:**

Using the mass formula, the integral is

$$\text{Mass} = \int_0^2 \int_0^x (x + 1) \, dy \, dx$$

$$= \int_0^2 (xy + y) \Big|_0^x \cdot dx$$

$$= \int_0^2 \left( x^2 + x \right) \cdot dx$$

$$= \frac{x^3}{3} + \frac{x^2}{2} \Big|_0^2$$

$$= 8/3 + 2 - 0 - 0 = 14/3 \text{ g}$$

◇

Once we have the ability to compute the mass of a plate from its dimensions and density, we can compute the coordinates of the center of mass of the plate using a type of averaging integral.

**Knowledge Box 3.11**

## Center of mass

*Suppose that a flat plate occupies a region R with a density function $\rho(x, y)$ defined on R. Then if*

$$M_x = \int \int_R y\rho(x, y) \cdot dA$$

*and*

$$M_y = \int \int_R x\rho(x, y) \cdot dA$$

*the center of mass of the plate is*

$$(\overline{x}, \overline{y}) = \left(\frac{M_y}{M}, \frac{M_x}{M}\right).$$

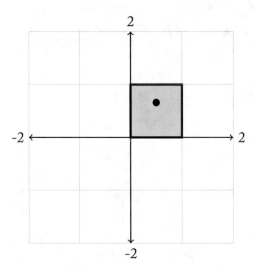

Figure 3.10: Center of mass of square region when $\rho(x, y) = 2y$ g/unit$^2$.

**Example 3.44**   Suppose that $R$ is the square region

$$0 \le x, y \le 1$$

and that $\rho(x, y) = 2y$ g/unit$^2$ as shown in Figure 3.10.

Find the center of mass.

**Solution:**

This problem requires three integrals.

$$M = \int_0^1 \int_0^1 2y \cdot dy \, dx \qquad M_x = \int_0^1 \int_0^1 y \cdot 2y \cdot dy \, dx \qquad M_y = \int_0^1 \int_0^1 x \cdot 2y \cdot dy \, dx$$

$$= \int_0^1 y^2 \Big|_0^1 dx \qquad\qquad = \int_0^1 \int_0^1 2y^2 \cdot dy \, dx \qquad\quad = \int_0^1 xy^2 \Big|_0^1 \cdot dx$$

$$= \int_0^1 (1 - 0) dx \qquad\qquad = \int_0^1 \frac{2}{3} y^3 \Big|_0^1 \cdot dx \qquad\qquad = \int_0^1 (x - 0) \cdot dx$$

$$= \int_0^1 dx \qquad\qquad\qquad = \int_0^1 \left( \frac{2}{3} - 0 \right) \cdot dx \qquad\qquad = \frac{1}{2} x^2 \Big|_0^1 = \frac{1}{2}$$

$$= x \Big|_0^1 \qquad\qquad\qquad = \int_0^1 \frac{2}{3} \cdot dx$$

$$= (1 - 0) = 1 \text{ gram} \qquad\quad = \frac{2}{3} x \Big|_0^1 = \frac{2}{3}$$

Now that we have the pieces we can use the formula for center of mass:

$$(\overline{x}, \overline{y}) = \left( \frac{1/2}{1}, \frac{2/3}{1} \right) = \left( \frac{1}{2}, \frac{2}{3} \right)$$

$\Diamond$

# PROBLEMS

**Problem 3.45**   Find the integral of each of the following functions over the specified region.

1. The function $f(x, y) = x + y^2$ on the strip

$$0 \leq x \leq 4 \quad 0 \leq y \leq 1.$$

2. The function $g(x, y) = xy$ on the rectangle

$$1 \leq x \leq 3 \quad 1 \leq y \leq 2.$$

3. The function $h(x, y) = x^2 y + x y^2$ on the square

$$0 \leq x \leq 2 \quad 0 \leq y \leq 2.$$

4. The function $r(x, y) = 2x + 3y + 1$ on the region bounded by $x = 0$, $y = 1$, and $y = x$.

5. The function $s(x, y) = x^2 + y^2 + 1$ on the region bounded by the $x$ axis and the function $y = 4 - x^2$.

6. The function $q(x, y) = x + y$ on the region bounded by the curves $y = \sqrt{x}$ and $y = x^2$.

7. The function $a(x, y) = x^2$ on the region bounded by the curves $y = 2x$ and $y = x^2$.

8. The function $b(x, y) = y^2$ on the region bounded by the curves $y = \sqrt[3]{x}$ and $y = x$ for $x \geq 0$.

**Problem 3.46**   Sketch the regions from Problem 3.45.

**Problem 3.47**   Explain why a density function $\rho(x, y)$ can never be negative.

**Problem 3.48**   Find a region $R$ so that the integral over $R$ of $f(x) = x^2 + y^2$ is 6 units$^3$.

**Problem 3.49**   Find the square region $0 \leq x, y \leq a$ so that

$$\int \int_R (x^3 + y) \cdot dA$$

is 12 units$^3$.

**Problem 3.50**    Find the mass of the plate $0 \le x, y \le 3$ if

$$\rho(x, y) = y^2 + 1.$$

**Problem 3.51**    Find the center of mass of the region bounded by the $x$-axis, the $y$-axis, and the line $x + y = 4$ if the density function is

$$\rho(x, y) = y + 1.$$

**Problem 3.52**    Find the center of mass of the region bounded by $y = \sqrt{x}$ and $y = x^2$ if the density function is

$$\rho(x, y) = x + 2.$$

**Problem 3.53**    Find the center of mass of the region $0 \le x, y \le 1$ if the density function is

$$\rho(x, y) = (x + y)/2.$$

**Problem 3.54**    Find the volume under

$$f(x, y) = \sqrt{x^2 + y^2}$$

above the region $x^2 + y^2 \le 16$.

**Problem 3.55**    Find the volume under

$$f(x, y) = x^2 + y^2$$

above the region bounded by the petal curve $r = 2\cos(3\theta)$.

**Problem 3.56**    Find the volume under

$$f(x, y) = (x^2 + y^2)^{3/2}$$

above the region bounded by the petal curve $r = \cos(2\theta)$.

**Problem 3.57**   Find a plane $f(x, y)$ so that the area under the plane but over a circle of radius 2 centered at the origin is 16 units$^3$.

**Problem 3.58**   Derive the general formula for the volume over a rectangle and under a plane in a region where the plane has a positive $z$ value.

CHAPTER 4

# Sequences, Series, and Function Approximation

This chapter is quite different from the rest of the material in the *Fast Start* series. Calculus is mostly part of the mathematics of continuous functions, while the sequences and series we study in this chapter are a part of **discrete mathematics** – math that is broken up into individual pieces. Discrete math is about things you can count, rather than things you can measure. Since this is a calculus book, we will apply what we learn to understanding and increasing the power of our calculus. The ultimate goal of the chapter is a much deeper understanding of transcendental (non-polynomial) functions like $e^x$ or $\cos(x)$. We start, however, at the beginning.

## 4.1 SEQUENCES AND THE GEOMETRIC SERIES

A sequence is an infinite list of numbers. Sometimes we give a sequence by listing an obvious pattern:

$$S = 1, \frac{1}{2}, \frac{1}{3}, \frac{1}{4}, \ldots$$

with the ellipsis meaning "and so on." We also can use more formal set notation to specify a sequence with a formula:

$$S = \left\{ \frac{1}{n} : n = 1, 2, \ldots \right\}$$

Like a function having a limit at infinity, there is a notion of a sequence **converging**.

### Knowledge Box 4.1

**Definition of the convergence of a sequence to a limit**

*We call L the **limit of a sequence** $\{x_n : n = 1, 2, \ldots\}$ if, for each $\epsilon > 0$, there is a whole number N so that, whenever $n > N$, we have:*

$$|x_n - L| < \epsilon.$$

*A sequence that has a limit is said to **converge**.*

**Example 4.1**   Prove that the sequence

$$S = \left\{ \frac{1}{n} : n = 1, 2, \ldots \right\}$$

converges to zero.

**Solution:**

For $\epsilon > 0$ choose $N$ to be the smallest whole number greater than $\dfrac{1}{\epsilon}$. This makes $\dfrac{1}{N} < \epsilon$. Then if $n > N$ we have:

$$n > N$$

$$\frac{1}{n} < \frac{1}{N} \qquad\qquad \text{Reciprocals reverse inequalities}$$

$$\frac{1}{n} < \epsilon \qquad\qquad \text{Since } 1/N < \epsilon$$

$$\left| \frac{1}{n} - 0 \right| < \epsilon \qquad\qquad \text{Value on the left did not change}$$

Which satisfies the definition of the limit of the sequence being $L = 0$.

As always there are shortcuts that mean we only need to rely on the definition of the limits of sequences occasionally.

**Knowledge Box 4.2**

**Sequences drawn from functions**

*Suppose that we have a sequence*

$$S = \{x_n\} = \{f(n) : n = 1, 2, \ldots\}$$

*and that* $\lim\limits_{x \to \infty} f(x) = L$. *Then we may conclude that* $\lim\limits_{n \to \infty} x_n = L$.

**Example 4.2**   Suppose that we have a sequence

$$\left\{ x_n = \frac{1}{1 + n^2} : n = 0, 1, 2, \ldots \right\}$$

Find $\lim_{n \to \infty} x_n$.

**Solution:**

We already know $\lim_{x \to \infty} \dfrac{1}{1 + x^2} = 0$. So, using Knowledge Box 4.2, we have that

$$\lim_{n \to \infty} x_n = 0$$

The rule in Knowledge Box 4.2 is *not reversible*.

**Example 4.3**   Determine if the sequence

$$S = \{\cos(2\pi n) : n = 0, 1, 2, \ldots\}$$

converges.

**Solution:**

Since $\lim_{x \to \infty} \cos(2\pi x)$ jumps back and forth in the range $-1 \le y \le 1$, the function that we drew the sequence from does not have a limit.

Leaving that aside, examine a listing of the first several terms of the sequence:

$$\{1, 1, 1, 1, 1, 1, \ldots\}$$

This sequence obviously converges to $L = 1$.

Again: if the function has a limit, then so does the sequence. The reverse need not be true.

The next sequence resolution technique requires that we define several terms.

**Definition 4.1** *If a sequence $\{x_n\}$ has the property that $x_n \leq x_{n+1}$, then we say that the sequence is* **monotone increasing***.*

**Definition 4.2** *If a sequence $\{x_n\}$ has the property that $x_n \geq x_{n+1}$, then we say that the sequence is* **monotone decreasing***.*

**Definition 4.3** *If a sequence $\{x_n\}$ is either monotone increasing or monotone decreasing, then we say the sequence is* **monotone***.*

**Definition 4.4** *If a sequence $\{x_n\}$ has the property that*

$$x_n \leq C$$

*for all n and for some constant $C$, then we say the sequence is* **bounded above***.*

**Definition 4.5** *If a sequence $\{x_n\}$ has the property that*

$$x_n \geq C$$

*for all n and for some constant $C$, then we say the sequence is* **bounded below***.*

**Definition 4.6** *If a sequence $\{x_n\}$ is both bounded above and bounded below, then we say the sequence is* **bounded***.*

**Knowledge Box 4.3**

**Bounded monotone sequences**

*The following types of sequences all converge.*

- *A monotone increasing sequence that is bounded above.*
- *A monotone decreasing sequence that is bounded below.*
- *Any monotone bounded sequence.*

The reason that the sequences listed in Knowledge Box 4.3 converge is that they are required to move toward a bound of some sort without passing it. This is another useful shortcut for determining if a sequence converges, although this shortcut does not tell us the value of the limit.

**Example 4.4**   Demonstrate that the sequence

$$\left\{ x_n = \frac{n}{n+1} : n = 0, 1, 2, 3, \ldots \right\}$$

has a limit.

**Solution:**

While we could use Knowledge Box 4.2, let's use this example as a chance to demonstrate the technique from Knowledge Box 4.3.

First notice that this sequence is monotone increasing. To see this, notice that:

$$n^2 + 2n < n^2 + 2n + 1$$

$$n(n + 2) < (n + 1)^2$$

$$\frac{n}{n+1} < \frac{n+1}{n+2}$$

$$x_n < x_{n+1}$$

Since $n < n + 1$, we can deduce that $x_n = \dfrac{n}{n+1} < 1$.

So the sequence is bounded above. This permits us to use Knowledge Box 4.3 to deduce that the sequence has a limit.

$$\Diamond$$

It would have been easier to do Example 4.4 with Knowledge Box 4.2. But there will be times when the sequence is not drawn from a function when we have to use monotone sequence theory. The next Knowledge Box extends the reach of our ability to check sequences for convergence.

**Knowledge Box 4.4**

### Arithmetic combinations of sequences

*If $\{x_n\}$ has a limit of $L$, and $\{y_n\}$ has a limit of $M$, and if $a, b$ are constants, then:*

- $\displaystyle\lim_{n\to\infty} a x_n \pm b y_n = aL \pm bM$

- $\displaystyle\lim_{n\to\infty} x_n \cdot y_n = L \cdot M$

- $\displaystyle\lim_{n\to\infty} \frac{x_n}{y_n} = \frac{L}{M}$ *if $M \neq 0$.*

- $\displaystyle\lim_{n\to\infty} x_n^k = L^k$

**Example 4.5**    Find the limit of

$$S = \left\{ x_n = \frac{1}{n} + 3 \cdot \frac{n}{n+1} : n = 1, 2, 3, \ldots \right\}$$

**Solution:**

We already know the limit of $1/n$, as a series, is zero. Using Knowledge Box 4.2 it is easy to see that:

$$\lim_{n\to\infty} \frac{n}{n+1} = 1$$

Combining these results using the information in Knowledge Box 4.4 we get that the limit is:

$$L = 0 + 3 \cdot 1 = 3$$

This concludes our direct investigation of sequences. We now turn to using sequences as a tool to explore **series**. Where a sequence is an infinite list of numbers, a series is an infinite list of numbers *that you add up*. This may or may not result in a finite sum – and resolving that question requires sequence theory.

**Definition of series**

If $\{x_n : n = 0, 1, \ldots\}$ is a sequence, then

$$\sum_{n=0}^{\infty} x_n = x_0 + x_1 + \cdots + x_k + \cdots$$

is the corresponding **infinite series**. If we sum only finitely many terms we have a **finite series**.

**Example 4.6** Show that the following infinite series sums to 1.

$$\sum_{n=1}^{\infty} \frac{1}{2}^n = \frac{1}{2} + \frac{1}{4} + \frac{1}{8} + \cdots$$

**Solution:**

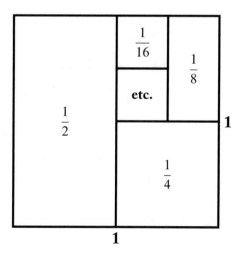

Figure 4.1: A unit square cut into pieces 1/2, 1/4, 1/8, . . .

Examine Figure 4.1. The figure divides a unit square (with area 1) into rectangles of size 1/2, 1/4, 1/8, and every other number in the series we are trying to sum. This constitutes a geometric demonstration that the series sums to 1.

◊

Clearly, finding a cool picture is not a general technique for demonstrating that a series has a sum. Proving no such picture exists when the series fails to have a sum is even more impossible – we need a more general theory.

---

**Knowledge Box 4.6**

### The sequence of partial sums of a series

*Suppose that* $S = \displaystyle\sum_{n=1}^{\infty} x_n$ *is a series. If we set*

$$p_n = \sum_{k=1}^{n} x_k,$$

*then* $\{p_n\}$ *is called the* **sequence of partial sums of** $S$. *We say that a series* **converges to a sum** $L$ *if and only if its sequence of partial sums has $L$ as a limit.*

---

**Example 4.7**    Find the sequence of partial sums of the series in Example 4.6 and compute its limit.

**Solution:**

Compute the first few members of the sequence of partial sums:

$$p_1 = 1/2$$

$$p_2 = 1/2 + 1/4 = 3/4$$

$$p_3 = 1/2 + 1/4 + 1/8 = 7/8$$

$$p_4 = 1/2 + 1/4 + 1/8 + 1/16 = 15/16$$

Which is a clear pattern, and it is easy to see

$$p_n = 1 - \frac{1}{2^n}$$

This gives us the sequence of partial sums. Computing the limit we get:

$$\lim_{n \to \infty} p_n = \lim_{n \to \infty} 1 - \frac{1}{2^n} = 1 - 0 = 1$$

So we get the same sum using this more formal approach.

$$\diamond$$

At this point we need to call forward an identity from *Fast Start Integral Calculus*:

$$1 + x + x^2 + \cdots x^n = \frac{x^{n+1} - 1}{x - 1} = \frac{1 - x^{n+1}}{1 - x}$$

This identity is one that is true for polynomials, but it also applies to summing a finite series. Additionally, this formula can be used as the partial sum of a particular type of infinite series – at least when its limit exists.

### Knowledge Box 4.7

#### Finite and infinite geometric series

*The polynomial identity in the text tells us, for a constant $a \neq 1$, that:*

$$\sum_{k=0}^{n} a^k = \frac{a^{n+1} - 1}{a - 1}.$$

*This is the* **finite geometric series formula.**

*If $|a| < 1$, then the limit of the finite series gives us the* **infinite geometric series formula:**

$$\sum_{n=0}^{\infty} a^n = \frac{1}{1 - a}.$$

*Applying Knowledge Box 4.4 we also get that:*

$$\sum_{n=0}^{\infty} c \cdot a^n = \frac{c}{1 - a}$$

*for a constant $c$. The number $a$ is called the* **ratio** *of the geometric series.*

**Example 4.8**    Compute

$$\sum_{n=0}^{\infty} \frac{3}{5^n}$$

**Solution:**

This is a geometric series with ratio $a = 1/5$. It has a leading constant $c = 3$. Applying the appropriate formula we see that:

$$\sum_{n=0}^{\infty} \frac{3}{5^n} = \frac{3}{1 - 1/5} = \frac{3}{4/5} = \frac{15}{4}$$

# PROBLEMS

**Problem 4.9**    Prove formally, using the definition of the limit of a sequence, that

$$\{\cos(2\pi n) : n = 0, 1, 2, \ldots\}$$

converges to 1.

**Problem 4.10**    Prove formally, using the definition of the limit of a sequence, that

$$\left\{ \frac{1}{n^2} : n = 1, 2, 3, \ldots \right\}$$

converges to 0.

**Problem 4.11**    Prove formally, using the definition of the limit of a sequence, that

$$\left\{ \frac{n}{n+1} : n = 0, 1, 2, \ldots \right\}$$

converges to 1.

**Problem 4.12**    Compute the limit of each of the following sequences or give a reason why the limit does not exist. Assume $n = 1, 2, \ldots$

1. $\{x_n = \tan^{-1}(n)\}$

2. $\left\{y_n = \sin\left(\dfrac{\pi}{2}n\right)\right\}$

3. $\left\{z_n = \dfrac{n^2}{n+1}\right\}$

4. $\{y_n = \sin(\pi n)\}$

5. $\left\{z_n = \dfrac{3n^2}{n^2+1}\right\}$

6. $\left\{z_n = \dfrac{\cos(n)}{n+1}\right\}$

**Problem 4.13**    Do the calculation to prove the infinite geometric series formula from the finite one: see Knowledge Box 4.7.

**Problem 4.14**    Compute the following sums or give a reason they fail to exist.

1. $\displaystyle\sum_{k=0}^{20} 1.2^k$

2. $\displaystyle\sum_{n=0}^{\infty} \left(\frac{1}{3}\right)^n$

3. $\displaystyle\sum_{n=0}^{\infty} 2 \cdot \left(\frac{1}{7}\right)^n$

4. $\displaystyle\sum_{n=0}^{\infty} \left(\frac{-1}{4}\right)^n$

5. $\displaystyle\sum_{n=0}^{\infty} 3\left(\frac{3}{2}\right)^n$

6. $\displaystyle\sum_{n=0}^{\infty} 0.05^n$

7. $\displaystyle\sum_{n=0}^{\infty} 112\,(0.065)^n$

8. $\displaystyle\sum_{n=0}^{\infty} 2 \cdot (-1)^n$

**Problem 4.15**    Compute the following sums or give a reason they fail to exist.

1. $\displaystyle\sum_{k=12}^{24} 3^n$

2. $\displaystyle\sum_{k=5}^{15} 2^n$

3. $\displaystyle\sum_{k=3}^{30} 1.2^n$

4. $\displaystyle\sum_{k=10}^{20} 4.5^n$

5. $\displaystyle\sum_{k=90}^{100} 1.1^n$

6. $\displaystyle\sum_{k=1}^{22} 7^n$

**Problem 4.16**    A swinging pendulum is losing energy. Its first swing is 2 m long, and each after that is 0.9985 times as long as the one before it. Estimate the total distance traveled by the pendulum.

**Problem 4.17**   A ball is dropped from a height of 8 m. If each bounce is 3/4 the height of the one before it, estimate the total vertical distance traveled by the ball.

**Problem 4.18**   A rod is initially displaced 2.1 mm from equilibrium and undergoes damped vibration with a decay in the length of each subsequent swing of 0.937. Find the total vertical distance traveled by the end of the rod.

**Problem 4.19**   A ball is dropped from a height of 4m. If each bounce is 0.86 times the height of the one before it, estimate the total vertical distance traveled by the ball.

**Problem 4.20**   A rod is initially displaced 3cm from equilibrium and undergoes damped vibration with a decay in the length of each subsequent swing of 0.987. Find the total vertical distance traveled by the end of the rod.

**Problem 4.21**   A very orderly and goes first north then east over an over. If the distances it travels before turning go 1, 1/3, 1/9, 1/27, and so on, what the is distance from its starting point that it approaches as it travels farther and farther?

## 4.2   SERIES CONVERGENCE TESTS

In this section we will develop a number of tests to determine if a series converges. At present, we know that an infinite geometric series with a ratio of $a$ with $|a| < 1$ will converge, but not much else. We begin with a motivating example.

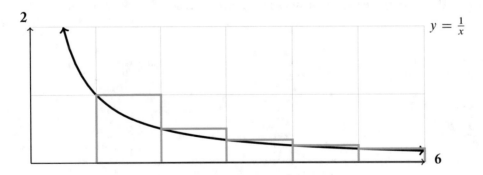

Figure 4.2: Shown is a portion of the graph of $y = 1/x$ and a sequence of rectangles of width 1 and height $1/n$ for $n = 1, 2, 3, \ldots$.

**Example 4.22**    Determine if the series $\sum\limits_{n=1}^{\infty} \dfrac{1}{n}$ has a finite or infinite sum.

**Solution:**

Examine Figure 4.2. This shows that a series of rectangles with area $1, 1/2, 1/3, \ldots$ have a larger area than $\int_{1}^{\infty} \dfrac{dx}{x}$ because the area under the curve is strictly smaller than the sum of the areas of the rectangles. Computing the integral:

$$\int_{1}^{\infty} \frac{dx}{x} = \lim_{a \to \infty} \int_{1}^{a} \frac{dx}{x}$$

$$= \lim_{a \to \infty} \ln(x)\Big|_{1}^{a}$$

$$= \lim_{a \to \infty} \ln(a) - \ln(1)$$

$$= \infty$$

shows us that: $\sum\limits_{n=1}^{\infty} \dfrac{1}{n} \geq \infty$, and we conclude the sum is infinite.

$\Diamond$

The series

$$\sum_{n=1}^{\infty} \frac{1}{n}$$

is sufficiently important that it has its own name: the **harmonic series**.

**Definition 4.7** *When considering*

$$\sum_{n=1}^{\infty} x_n$$

*we call $x_n$ the* **general term** *of the series.*

Example 4.22 is another example of proving something about a series sum by drawing a clever picture. An interesting technique, but, as before, we need more general tools. One such tool involves taking the limit of the sequence generating the series. If it does not approach zero, there is no hope that the infinite sum converges.

---

**Knowledge Box 4.8**

**The divergence test**

*If* $\lim_{n \to \infty} x_n \neq 0$ *then*

$$\sum_{n=0}^{\infty} x_n$$

*does not have a finite value.*

---

It is important to note that the divergence test is uni-directional. If the general term of a series *does* go to zero, that tells you exactly nothing about the behavior of the sum of the series.

**Example 4.23**   Show that

$$\sum_{n=0}^{\infty} \frac{n}{n+1}$$

diverges (has an infinite sum).

**Solution:**

Since

$$\lim_{n \to \infty} \frac{n}{n+1} = 1,$$

the series in question diverges by the divergence test. Colloquially, we are adding up an infinite number of terms that are approaching one – so the resulting sum is infinite.

When you can use it, the divergence test is often short and sweet. The next test is the formal version of the test we used in Example 4.22.

**Knowledge Box 4.9**

## The integral test

*Suppose that $f(x)$ is a positive, decreasing function on $[0, \infty)$ and that $x_n = f(n)$. Then,*

$$\sum_{n=1}^{\infty} x_n \ and \ \int_a^{\infty} f(x)dx$$

*both converge or both diverge for any finite $a \geq 1$.*

**Example 4.24**    Show that

$$\sum_{n=1}^{\infty} \frac{1}{n^2}$$

is finite.

**Solution:**

Use the integral test.

$$\int_1^{\infty} \frac{dx}{x^2} = \lim_{a \to \infty} \int_1^a \frac{dx}{x^2}$$

$$= \lim_{a \to \infty} \frac{-1}{x} \Big|_1^a$$

$$= \lim_{a \to \infty} \frac{-1}{a} - (-1)$$

$$= 0 + 1 = 1$$

Since the improper integral is finite, so is the sum. Note that this *does not* tell us the value of the sum.

◊

The fact that the series $\sum_{n=1}^{\infty} \frac{1}{n}$ diverges but $\sum_{n=1}^{\infty} \frac{1}{n^2}$ converges motivates our next test.

**Definition 4.8** *A p-series is a series of the form*

$$\sum_{n=1}^{\infty} \frac{1}{n^p}$$

*where p is a constant.*

**Knowledge Box 4.10**

**The $p$-series test**

*The p-series*

$$\sum_{n=1}^{\infty} \frac{1}{n^p}$$

*converges if p > 1 and diverges if p ≤ 1.*

The series convergence test in Knowledge Box 4.10 follows directly from the integral test – something you are asked to verify in the homework.

**Example 4.25**   Determine if

$$\sum_{n=1}^{\infty} \frac{1}{n^{\pi}}$$

converges to a finite number or diverges.

**Solution:**

This series has the form of a $p$-series with $p = \pi$. Since $\pi > 1$, we conclude the series converges.

$$\Diamond$$

**Example 4.26**   Determine if

$$\sum_{n=1}^{\infty} \frac{1}{\sqrt{n}}$$

converges to a finite number or diverges.

**Solution:**

This series has the form of a *p*-series with $p = 1/2$. Since $1/2 \leq 1$, we conclude the series diverges.

$$\Diamond$$

The next two tests leverage series we already understand to resolve even more series. The tests depend on various forms of comparison of a series under test to a series with known convergence or divergence behavior.

### Knowledge Box 4.11

#### The comparison tests

- If $\displaystyle\sum_{n=0}^{\infty} x_n$ *converges and* $0 \leq y_n \leq x_n$ *for all n, we have that* $\displaystyle\sum_{n=0}^{\infty} y_n$ *also converges.*

- If $\displaystyle\sum_{n=0}^{\infty} x_n$ *diverges and* $y_n \geq x_n \geq 0$ *for all n, we have that* $\displaystyle\sum_{n=0}^{\infty} y_n$ *also diverges.*

The colloquial versions of the comparison tests are pretty easy to believe. The first says, of a series with positive general terms, that if it is smaller, term by term, than another series with a finite sum, then it has a finite sum. The second reverses that: saying that, if a series is larger, term by term, than another series with an infinite sum, then it has an infinite sum.

**Example 4.27**   Determine if

$$\sum_{n=1}^{\infty} \frac{1}{n/3}$$

converges or diverges.

**Solution:**

Notice that, for $n \geq 1$,

$$n \geq n/3$$

$$\frac{1}{n} \leq \frac{1}{n/3}$$

So $0 \leq \dfrac{1}{n} \leq \dfrac{1}{n/3}$, which permits us to deduce that the series in the example diverges by comparison to the harmonic series, which is known to diverge.

$\Diamond$

**Example 4.28**    Determine if

$$\sum_{n=0}^{\infty} \frac{1}{2^n + 5}$$

converges or diverges.

**Solution:**

The key to using the comparison test is to find a good known series to compare to. In this case the geometric series with ratio 1/2 is natural.

$$2^n \leq 2^n + 5$$

$$\frac{1}{2^n} \geq \frac{1}{2^n + 5}$$

$$\left(\frac{1}{2}\right)^n \geq \frac{1}{2^n + 5}$$

Which means that the general term of the convergent geometric series

$$\sum_{n=0}^{\infty} \left(\frac{1}{2}\right)^n$$

is greater than the general term of our target series. Since all terms are positive, this tells us that we may conclude the target series converges, by comparison.

$\Diamond$

The next test is one of the most all-around useful tests. It permits us to resolve any series drawn from a rational function, for example, by checking to see which $p$-series a rational function is most like.

**Knowledge Box 4.12**

**The limit comparison test**

*Suppose that*

$$\sum_{n=0}^{\infty} x_n \text{ and } \sum_{n=0}^{\infty} y_n$$

*are series, and that*

$$\lim_{n\to\infty} \frac{x_n}{y_n} = C$$

*where $0 < |C| < \infty$ is a constant. Then, either both series converge or both series diverge.*

**Example 4.29**   Determine if

$$\sum_{n=1}^{\infty} \frac{n+5}{n^3+1}$$

converges or diverges.

**Solution:**

Perform limit comparison to

$$\sum_{n=1}^{\infty} \frac{1}{n^2}$$

$$\lim \frac{x_n}{y_n} = \lim_{n\to\infty} \frac{\frac{n+5}{n^3+1}}{\frac{1}{n^2}} = \lim_{n\to\infty} \frac{n+5}{n^3+1} \cdot \frac{n^2}{1} = \lim_{n\to\infty} \frac{n^3+5n^2}{n^3+1} = 1$$

Since $0 < 1 < \infty$, we can conclude that $\sum_{n=1}^{\infty} \frac{n+5}{n^3+1}$ converges, because $\sum_{n=1}^{\infty} \frac{1}{n^2}$ is a convergent $p$-series.

◊

Notice that choosing a $p$-series with $p = 2$ was exactly what was needed to make the ratio of the general terms of the series be something that had a finite, non-zero limit. The limit comparison test is useful for putting rigour into the intuition that one series is "like" another.

The next test uses the fact that if you have a sum of positive terms that is finite, making some or all of those terms negative cannot take the sum farther from zero than the original finite sum – leaving the sum finite.

**Knowledge Box 4.13**

**The absolute convergence test**

$$\text{If } \sum_{n=0}^{\infty} |x_n| \text{ converges, then so does } \sum_{n=0}^{\infty} x_n.$$

**Example 4.30**    Prove that

$$1 - \frac{1}{4} + \frac{1}{9} - \frac{1}{16} + \frac{1}{25} - \cdots$$

converges.

**Solution:**

The series in question is given as an obvious pattern – begin by pulling it into summation notation:

$$\sum_{n=1}^{\infty} \frac{(-1)^{n+1}}{n^2}$$

If we take the absolute value of the general terms of this series, we obtain the series:

$$\sum_{n=1}^{\infty} \frac{1}{n^2}$$

We know this to be a convergent $p$-series. We may deduce that the series converges.

◊

**Definition 4.9** *If* $\sum_{n=0}^{\infty} |x_n|$ *converges, then we say the series* $\sum_{n=0}^{\infty} x_n$ **converges in absolute value.**

This means we can restate Knowledge Box 4.13 as: "A series that converges in absolute value, converges."

The next test uses the fact that if you go up, then down by less, then up by even less, and so on, you end up a finite distance from your starting point.

**Knowledge Box 4.14**

**The alternating series test**

*Suppose that* $x_n$ *is a series such that* $x_n > x_{n+1} \geq 0$, *and suppose that* $\lim_{n \to \infty} x_n = 0$. *Then the series*

$$\sum_{n=0}^{\infty} (-1)^n \cdot x_n$$

*converges.*

**Example 4.31** Show that $\sum_{n=1}^{\infty} \dfrac{(-1)^n}{\sqrt{n}}$ converges.

**Solution:**

We already know that $\lim_{n \to \infty} \dfrac{1}{\sqrt{n}} = 0$, and the terms of the series clearly get smaller as $n$ increases. So, we may conclude this series converges by the alternating series test.

$$\diamond$$

The next two tests both check to see if a series is like a geometric series and deduce its convergence or divergence from that similarity.

**Knowledge Box 4.15**

**The ratio test**

*Suppose that* $\sum_{n=0}^{\infty} x_n$ *is a series. Compute*

$$r = \lim_{n \to \infty} \left| \frac{x_{n+1}}{x_n} \right|.$$

*Then:*

- *if* $r < 1$, *then the series converges,*

- *if* $r > 1$, *then the series diverges,*

- *if* $r = 1$, *then the test is inconclusive.*

**Definition 4.10** *The quantity* $n! = n \cdot (n-1) \cdot (n-2) \cdots 2 \cdot 1$ *is called* $n$ **factorial**.

*We define* $0! = 1$.

**Example 4.32**   Determine if $\sum_{n=0}^{\infty} \frac{1}{n!}$ converges or diverges.

**Solution:**

Use the ratio test.

$$\lim_{n \to \infty} \left| \frac{1/(n+1)!}{1/n!} \right| = \lim_{n \to \infty} \left| \frac{n!}{(n+1)!} \right|$$

$$= \lim_{n \to \infty} \frac{n(n-1) \cdots 2 \cdot 1}{(n+1)n(n-1) \cdots 2 \cdot 1}$$

$$= \lim_{n \to \infty} \frac{1}{n+1} = 0$$

Since $0 < 1$, we can deduce that the series converges by the ratio test.

◊

**Knowledge Box 4.16**

**The root test**

*Suppose that $\sum_{n=0}^{\infty} x_n$ is a series. Compute*

$$s = \lim_{n \to \infty} \sqrt[n]{|x_n|}.$$

*Then:*

- *if $s < 1$, then the series converges,*
- *if $s > 1$, then the series diverges,*
- *if $s = 1$, then the test is inconclusive.*

**Example 4.33**    Determine if $\sum_{n=0}^{\infty} \dfrac{1}{n^n}$ converges or diverges.

**Solution:**

Use the root test.

$$\lim_{n \to \infty} \sqrt[n]{\left|\frac{1}{n^n}\right|} = \lim_{n \to \infty} \frac{1}{n} = 0$$

Since $0 < 1$, we can deduce that the series converges by the root test.

The root test and, especially, the ratio test will get a big workout in the next section. This section contains nine tests for series convergence – many of which depend on knowing the convergence or divergence behavior of other series. This creates a mental space very similar to the "which integration method do I use?" issue that arose in *Fast Start Integral Calculus*. The method for dealing with this is the same here as it was there: practice, practice, practice.

It's also good to keep in mind, when you are searching for series to compare to, that the examples in this section may be used as examples for comparison. A list of series with known behaviors is an excellent resource and, given different learning styles, somewhat personal: you may want to maintain your own annotated list.

## 4.2.1   TAILS OF SEQUENCES

You may have noticed that we are a little careless with where we start the index of summation on our infinite series. This is because, while the value of a convergent series depends on every term, the convergence or divergence behavior does not.

**Definition 4.11** *If we take a sequence and make a new sequence by discarding a finite number of initial terms, the new sequence is a **tail** of the old sequence.*

**Knowledge Box 4.17**

**Tail convergence**

*A sequence converges if and only if all its tails converge.*

The practical effect of Knowledge Box 4.17 is that, if a few initial terms of a sequence are causing trouble, you may discard them and test the remainder of the sequence to determine convergence or divergence.

**Example 4.34**   Determine the convergence of the series:

$$\sum_{n=0}^{\infty} \frac{(-1)^n}{n^2 - 6n + 10}$$

**Solution:**

If we look at the first several terms of this series we get:

$$\frac{1}{10} - \frac{1}{5} + \frac{1}{2} - 1 + \frac{1}{2} - \frac{1}{5} + \frac{1}{10} - \frac{1}{17} + \cdots$$

The series alternates signs, getting larger in absolute value for the first four terms, but then getting smaller in absolute value for the remaining terms. This means that the alternating series test works *for the tail of the sequence starting at the fourth term*. Remember that the alternating series test requires that the terms shrink in absolute value.

We conclude that the series converges by the alternating series test applied to a tail of the sequence.

$$\Diamond$$

The next example shows a very special sort of series for which we can compute the exact value. Once you understand how these series work, you can construct many examples of them. You may want to review *partial fractions* from *Fast Start Integral Calculus*.

**Example 4.35**    Prove that the sequence

$$\sum_{n=0}^{\infty} \frac{1}{n^2 + 3n + 2}$$

converges to exactly 1.

**Solution:**

We can see, by limit comparison to $\sum_{n=1}^{\infty} \frac{1}{n^2}$, a convergent $p$-series, that this series converges. Knowing the exact value of the sum is another matter. That will require a little algebra.

$$\sum_{n=0}^{\infty} \frac{1}{n^2 + 3n + 2} = \sum_{n=0}^{\infty} \frac{1}{(n+1)(n+2)}$$

$$= \sum_{n=0}^{\infty} \left( \frac{A}{n+1} + \frac{B}{n+2} \right) \qquad \text{-partial fractions}$$

$$= \sum_{n=0}^{\infty} \left( \frac{1}{n+1} - \frac{1}{n+2} \right)$$

$$= 1 - 1/2 + 1/2 - 1/3 + 1/3 - 1/4 + 1/4 - 1/5 + \cdots$$

$$= 1 + 0 + 0 + 0 + 0 + \cdots$$

$$= 1$$

$$\Diamond$$

The series in Example 4.35 is called a *telescoping series*, in analogy to the way a small telescope or spyglass collapses for storage. In essence, the series is composed of a positive and a negative series that cancel out all but one term.

**Knowledge Box 4.18**

---

**Telescoping series**

If $b_n = a_n - a_{n+1}$ *then*

$$\sum_{n=0}^{\infty} b_n = a_0.$$

*When expanded in terms of the $a_k$ values, everything except $a_0$ cancels out.*

---

Trace out the information in Knowledge Box 4.18 for Example 4.35.

The next Knowledge Box gives a guide for choosing a convergence test. As with choosing the best method for integration, the only real way to get better at choosing the correct convergence test is to work examples. Lots of examples.

**Knowledge Box 4.19**

---

**Choosing a convergence test**

1. *If the general term of the series fails to converge to zero, then it diverges by the divergence test. Remember that this does not work in reverse – if the general term converges to zero, anything might happen.*

2. *If the series is a geometric series or a p-series, use the tests for those series.*

3. *If the general term of the series is $a_n = f(n)$ for a function $f(x)$ that you can integrate, try the integral test.*

4. *A positive series that is, term-by-term, no larger than a series that converges, converges (comparison test) – look for this.*

---

*continued*

**Knowledge Box 4.20**

### Choosing a convergence test—Continued

5. *Similarly, a positive series that is, term-by-term, no smaller than a divergent series diverges, again by the comparison test.*

6. *If you have a series that looks like a series you know how to deal with, try the limit comparison test. This is useful for things like slightly modified p series or letting you use simpler integral tests.*

7. *You can often set up a comparison test or limit comparison test by doing algebra or arithmetic to the general term of a series.*

8. *Remember that if the term-by-term absolute value of a series converges, then the series converges. This is the absolute convergence test.*

9. *If the terms of a series alternate in sign, look at the alternating series test.*

10. *If you can take the limit of adjacent terms of a series in a reasonable way, then the ratio test is a possibility.*

11. *If you can take the nth root of the general term of a series in a reasonable way, the root test is a possibility.*

12. *If some finite number of initial terms of a series are preventing you from using a test, tail convergence says you can ignore them and then do your test.*

13. *Unless you're taking a test or quiz, asking for advise and suggestions is not the worst possible option.*

# PROBLEMS

**Problem 4.36**   For each of the following series, determine if the series converges or diverges. State the name of the test you are using.

1. $\displaystyle\sum_{n=1}^{\infty} \frac{1}{n^{e}}$

2. $\displaystyle\sum_{n=0}^{\infty} \frac{n+1}{n^2+1}$

3. $\displaystyle\sum_{n=0}^{\infty} \frac{n+1}{5n+4}$

4. $\displaystyle\sum_{n=1}^{\infty} \ln\left(\frac{1}{n^2}\right)$

5. $\displaystyle\sum_{n=0}^{\infty} \frac{2^n+1}{3^n+1}$

6. $\displaystyle\sum_{n=0}^{\infty} \frac{\sqrt{n}}{n^2+1}$

**Problem 4.37**   For each of the following series, determine if the series converges or diverges. State the name of the test you are using.

1. $\displaystyle\sum_{n=0}^{\infty} \frac{n^2}{2^n+1}$

2. $\displaystyle\sum_{n=0}^{\infty} 0.0462^n$

3. $\displaystyle\sum_{n=2}^{\infty} \frac{\sin(n)}{n\sqrt{2}}$

4. $\displaystyle\sum_{n=1}^{\infty} \frac{1}{\sqrt[3]{n}}$

5. $\displaystyle\sum_{n=1}^{\infty} \frac{(-1)^n}{\sqrt[3]{n}}$

6. $\displaystyle\sum_{n=0}^{\infty} \frac{1}{(2n)!}$

**Problem 4.38**   For each of the following series, determine if the series converges or diverges. State the name of the test you are using.

1. $\displaystyle\sum_{n=1}^{\infty} \frac{1}{n^{n/2}}$

2. $\displaystyle\sum_{n=1}^{\infty} \frac{5^n}{n^{n/2}}$

3. $\displaystyle\sum_{n=1}^{\infty} e^{-n}$

4. $\displaystyle\sum_{n=1}^{\infty} \frac{e^n}{\pi^n}$

5. $\displaystyle\sum_{n=1}^{\infty} \frac{e^{2n}}{\pi^n}$

6. $\displaystyle\sum_{n=1}^{\infty} e^{3+n-n^2}$

**Problem 4.39**   Give an example to demonstrate that the divergence test does not work in reverse, i.e., a sequence whose general term goes to zero but whose sum is infinite.

**Problem 4.40**   Use the integral test to prove that the $p$-test works.

**Problem 4.41**   Suppose that $p(x)$ is a polynomial. Use the integral test to demonstrate that

$$\sum_{n=0}^{\infty} p(n)e^{-n}$$

converges.

**Problem 4.42**   Suppose that $q(x)$ is a polynomial with exactly three roots, all of which are negative real numbers. Demonstrate that

$$\sum_{n=0}^{\infty} \frac{1}{q(n)}$$

converges.

**Problem 4.43**   If $x_n$ and $y_n$ are the general terms of a convergent series, then $x_n + y_n$ are as well. This requires only simple algebra. What is startling is that the reverse is not true. Find an example of $a_n = x_n + y_n$ so that

$$\sum_{n=1}^{\infty} a_n$$

converges, but neither of

$$\sum_{n=1}^{\infty} x_n \text{ or } \sum_{n=1}^{\infty} y_n$$

converge.

**Problem 4.44**   Show that, when you apply the ratio test to a geometric series, the limit that appears in the test is the ratio of the series.

**Problem 4.45**   Show that, when you apply the root test to a geometric series, the limit that appears in the test is the ratio of the series.

**Problem 4.46**   Suppose $r > 1$. Prove that

$$\sum_{n=0}^{\infty} \frac{n^k}{r^n}$$

converges when $k$ is an integer $\geq 1$.

**Problem 4.47**   Suppose that we are testing the convergence of a series

$$\sum_{n=0}^{\infty} \frac{p(n)}{q(n)}$$

where $p(x)$ and $q(x)$ are polynomials. If we add the assumption that $q(x)$ has no roots at any of the values of $n$ involved in the sum, explain in terms of the degrees of the polynomials when the series converges.

**Problem 4.48**   For each of the following series, determine if the series converges or diverges. State the name of the test you are using.

1. $\displaystyle\sum_{n=0}^{\infty} \frac{n^3}{(1+n)(2+n)(3+n)(4+n)}$

4. $\displaystyle\sum_{n=0}^{\infty} \frac{\sqrt{n}}{n^2+1}$

2. $\displaystyle\sum_{n=0}^{\infty} \frac{n^2}{(1+n)(2+n)(3+n)(4+n)}$

5. $\displaystyle\sum_{n=0}^{\infty} 1.25^{-n}$

3. $\displaystyle\sum_{n=0}^{\infty} \frac{n^5}{e^n}$

6. $\displaystyle\sum_{n=0}^{\infty} e^{\pi-n}$

**Problem 4.49**   Compute exactly: $\displaystyle\sum_{n=1}^{\infty} \frac{1}{n^2+n}$

**Problem 4.50**   Compute exactly: $\displaystyle\sum_{n=1}^{\infty} \frac{1}{n^2+2n}$

**Problem 4.51**   Compute exactly: $\displaystyle\sum_{n=1}^{\infty} \frac{1}{n^2+5n}$

**Problem 4.52**   Demonstrate convergence of the series $\displaystyle\sum_{n=1}^{\infty} \frac{\sin(n)}{n^2+n}$

**Problem 4.53**   Compute exactly: $\displaystyle\sum_{n=0}^{\infty} \frac{1}{4n^2+8n+3}$

## 4.3   POWER SERIES

In this section we study series again. The good news is that we do not have any additional convergence tests. The bad news is that these series will have variables in them.

**Definition 4.12** *A* **power series** *is a series of the form:*

$$\sum_{n=0}^{\infty} a_n x^n$$

In a way, a power series is actually an infinite number of different ordinary series, one for each value of $x$ you could substitute into it. The goal of this section will be: given a power series, find values of $x$ which cause it to converge.

### Knowledge Box 4.21

**The radius of convergence of a power series**

*The power series*

$$\sum_{n=0}^{\infty} a_n x^n$$

*converges in one of three ways:*

1. *Only at $x = 0$.*

2. *For all $|x| < r$ and possibly at $x = \pm r$.*

3. *For all $x$.*

*The number $r$ is the* **radius of convergence** *of the power series. In the first case above, we say the radius of convergence is zero; in the third, we say the radius of convergence is infinite.*

**Definition 4.13** *The* **interval of convergence** *of a power series is the set of all $x$ where it converges.*

Knowledge Box 4.21 implies that the interval of convergence of a power series is one of $[0, 0]$, $(-r, r)$, $[-r, r)$, $(-r, r]$, $[-r, r]$, or $(-\infty, \infty)$. The results with an $r$ in them occur in the case where $0 < r < \infty$. Once we have the radius of convergence, in the case where $r$ is positive and finite, we determine the interval of convergence by checking the behavior of the series when we set $x = \pm r$.

**Example 4.54**   Find the radius and interval of convergence of:

$$\sum_{n=0}^{\infty} x^n$$

**Solution:**

We start by trying to determine the radius of convergence.

Use the ratio test:

$$\lim_{n \to \infty} \left| \frac{x^{n+1}}{x^n} \right| = \lim_{n \to \infty} |x| = |x|$$

This is true because $x$ does not depend on $n$.

The series thus converges when $|x| < 1$, meaning we have convergence for sure when $-1 < x < 1$.

This also means the radius of convergence is $r = 1$.

We now need to check $x = \pm 1$ to determine the interval of convergence.

These values both yield non-converging geometric series:

$$\sum_{n=0}^{\infty} (-1)^n \text{ and } \sum_{n=0}^{\infty} 1$$

So the potential endpoints of the interval of convergence are *not* part of the interval of convergence. This means that the interval of convergence is $(-1, 1)$.

$\Diamond$

**Example 4.55**   Find the radius of convergence of:

$$\sum_{n=0}^{\infty} \frac{x^n}{2^n}$$

**Solution:**

Again, use the ratio test.

$$\lim_{n \to \infty} \left| \frac{a_{n+1}}{a_n} \right| = \lim_{n \to \infty} \left| \frac{x^{n+1}/2^{n+1}}{x^n/2^n} \right|$$

$$= \lim_{n \to \infty} \left| \frac{x^{n+1}}{2^{n+1}} \cdot \frac{2^n}{x^n} \right|$$

$$= \lim_{n \to \infty} \left| \frac{x}{2} \right| = \frac{|x|}{2}$$

So the series converges when:

$$-1 < \frac{x}{2} < 1$$

$$-2 < x < 2$$

The radius of convergence is $r = 2$.

◇

**Example 4.56**   Find the radius of convergence of the series:

$$\sum_{n=0}^{\infty} \frac{x^n}{n!}$$

**Solution:**

Like before:

$$\lim_{n\to\infty}\left|\frac{a_{n+1}}{a_n}\right| = \lim_{n\to\infty}\left|\frac{x^{n+1}/(n+1)!}{x^n/n!}\right|$$

$$= \lim_{n\to\infty}\left|\frac{x}{n+1}\right|$$

$$= |x|\lim_{n\to\infty}\left|\frac{1}{n+1}\right|$$

$$= |x|\cdot 0$$

$$= 0$$

Since $0 < 1$ for all values of $x$, this power series converges everywhere and the radius of convergence is infinite.

◇

**Example 4.57**   Find the radius of convergence of the series:

$$\sum_{n=0}^{\infty}\frac{x^n}{n^n}$$

**Solution:**

This time use the root test.

$$\lim_{n\to\infty}\sqrt[n]{\left|\frac{x^n}{n^n}\right|} = \lim_{n\to\infty}\left|\frac{x}{n}\right|$$

$$= |x|\cdot\lim_{n\to\infty}\frac{1}{n} = 0$$

Which tells us that, as $0 < 1$ for all $x$, that this sequence converges everywhere.

◇

**Example 4.58**  Find the interval of convergence of the series:

$$\sum_{n=1}^{\infty} \frac{x^n}{n^2}$$

**Solution:**

Another natural job for the ratio test.

$$\lim_{n \to \infty} \left| \frac{a_{n+1}}{a_n} \right| = \lim_{n \to \infty} \left| \frac{x^{n+1}/(n+1)^2}{x^n/n^2} \right|$$

$$= \lim_{n \to \infty} \left| \frac{n^2}{(n+1)^2} \cdot x \right|$$

$$= |x| \cdot \lim_{n \to \infty} \frac{n^2}{n^2 + 2n + 1}$$

$$= |x| \cdot 1 = |x|$$

So $-1 < x < 1$, and the radius of convergence is $r = 1$. To find the interval of convergence we need to check the ends of the interval. These are of the form

$$\sum_{n=1}^{\infty} \frac{(\pm 1)^n}{n^2}$$

The absolute value of each of these is a $p$-series with $p = 2$. Both endpoints correspond to series that converge in absolute value – meaning they both converge. This makes the interval of convergence $[-1, 1]$.

◊

### 4.3.1  USING CALCULUS TO FIND SERIES

It is sometimes useful to represent functions as power series. The geometric series formula (Knowledge Box 4.7) does this, for example, for the function $f(x) = \dfrac{1}{1-x}$. We can use calculus to derive power series for other functions.

**Example 4.59**   If $|x| < 1$, then the geometric series formula tells us that:

$$\sum_{n=0}^{\infty} x^n = \frac{1}{1-x}$$

If we plug $-x$ into this identity we find that:

$$\sum_{n=0}^{\infty} (-x)^n = \frac{1}{1-(-x)}$$

$$\sum_{n=0}^{\infty} (-1)^n x^n = \frac{1}{1+x}$$

or

$$\frac{1}{x+1} = 1 - x + x^2 - x^3 + x^4 - \cdots$$

Integrate both sides and we get:

$$\ln(x+1) + C = x - \frac{1}{2}x^2 + \frac{1}{3}x^3 - \frac{1}{4}x^4 + \frac{1}{5}x^5 - \cdots$$

Plug in $x = 0$

$$\ln(1) + C = 0$$

$$C = 0$$

Which means, at least when $|x| < 1$,

$$\ln(x+1) = \sum_{n=1}^{\infty} \frac{(-1)^{n+1} x^n}{n}$$

$\Diamond$

This shows that we can find power series that, when they converge, are equal to familiar transcendental functions. The next section gives another technique for doing this, to be used when the sequences you already know don't give you enough power. Let's encode this as a Knowledge Box.

### Knowledge Box 4.22

**Using calculus to modify power series**

Suppose that $f(x) = \sum_{n=0}^{\infty} a_n x^n$. Then:

$$\int f(x) \cdot dx = \sum_{n=0}^{\infty} \frac{a_n}{n+1} x^{n+1} \text{ and } f'(x) = \sum_{n=1}^{\infty} n \cdot a_n x^{n-1}.$$

**Example 4.60** Find a power series for $\tan^{-1}(x)$.

**Solution:**

We already have seen that:

$$\frac{1}{u+1} = 1 - u + u^2 - u^3 + u^4 - \cdots$$

If we substitute $u = x^2$ we obtain:

$$\frac{1}{x^2+1} = 1 - x^2 + x^4 - x^6 + x^8 - \cdots$$

Integrate and we get:

$$\tan^{-1}(x) + C = x - \frac{1}{3}x^3 + \frac{1}{5}x^5 - \frac{1}{7}x^7 + \frac{1}{9}x^9 - \cdots$$

Substitute in $x = 0$ and we get $C = 0$. So we obtain the power series:

$$\tan^{-1}(x) = \sum_{n=0}^{\infty} \frac{(-1)^n x^{2n+1}}{2n+1}$$

$\Diamond$

**Example 4.61** What is the interval of convergence for the series for

$$f(x) = \tan^{-1}(x)$$

found in Example 4.60?

**Solution:**

Apply the ratio test:

$$\lim_{n\to\infty} \left| \frac{a_{n+1}}{a_n} \right| = \lim_{n\to\infty} \left| \frac{x^{2n+3}/(2n+3)}{x^{2n+1}/(2n+1)} \right|$$

$$= \lim_{n\to\infty} \left| x^2 \frac{2n+1}{2n+3} \right|$$

$$= |x^2| \lim_{n\to\infty} \frac{2n+1}{2n+3}$$

$$= |x^2| \cdot 1 = x^2$$

So $x^2 < 1$ when $-1 < x < 1$, making the radius of convergence $r = 1$.

Now check the endpoints $x = \pm 1$.

If $x = -1$ the resulting series is:

$$\sum_{n=0}^{n} \frac{(-1)^n (-1)^{2n+1}}{2n+1} = \sum_{n=0}^{n} \frac{(-1)^{n+1}}{2n+1}$$

which converges by the alternating series test.

If $x = 1$ we get:

$$\sum_{n=0}^{n} \frac{(-1)^n}{2n+1}$$

which also converges by the alternating series test.

The interval of convergence is thus $[-1, 1]$.

It is possible to find a power series by just using algebra on a known series.

**Example 4.62**   Find a power series for:

$$f(x) = \frac{x^2}{1 - x^2}$$

**Solution:**

We start with the known form:

$$\frac{1}{1 - u} = 1 + u + u^2 + u^3 + u^4 + \cdots$$

Substitute in $u = x^2$ and we get:

$$\frac{1}{1 - x^2} = 1 + x^2 + x^4 + x^6 + x^8 + \cdots$$

Now multiply both sides by $x^2$ and we get:

$$\frac{x^2}{1 - x^2} = x^2 + x^4 + x^6 + x^8 + x^{10} + \cdots$$

So

$$f(x) = \sum_{n=0}^{\infty} x^{2n+2}$$

◇

**Example 4.63**    Find a power series for:

$$g(x) = \ln(x^2 + 1)$$

**Solution:**

Start with the known result for $\dfrac{1}{1+x^2}$.

$$\frac{1}{1+x^2} = 1 - x^2 + x^4 - x^6 + x^8 - \cdots$$

$$\frac{2x}{1+x^2} = 2x - 2x^3 + 2x^5 - 2x^7 + 2x^9 - \cdots$$

$$\int \frac{2x}{1+x^2} \cdot dx = \int \left(2x - 2x^3 + 2x^5 - 2x^7 + 2x^9 - \cdots\right) dx$$

$$\ln(x^2+1) + C = 2\left(\frac{1}{2}x^2 - \frac{1}{4}x^4 + \frac{1}{6}x^6 - \frac{1}{8}x^8 + \frac{1}{10}x^{10} - \cdots\right)$$

Set $x = 0$ and we get $C = 0$.

So: $\ln(x^2 + 1) = \displaystyle\sum_{n=0}^{\infty} \frac{2 \cdot (-1)^n x^{2n+2}}{2n+2}$

◇

An integral that is both important in statistics, because it is related to the normal distribution, and famous for not having a closed form solution is:

$$\int e^{-x^2/2} \cdot dx$$

The next example shows why power series are useful – it is easy to get a power series for this integral.

**Example 4.64**   Find a power series for

$$F(x) = \int e^{-x^2/2} \cdot dx$$

**Solution:**

$$e^x = \sum_{n=0}^{\infty} \frac{x^n}{n!} \qquad\qquad \text{This will be shown in Section 4.4}$$

$$e^{-x^2/2} = \sum_{n=0}^{\infty} \frac{\left(-x^2/2\right)^n}{n!}$$

$$e^{-x^2/2} = \sum_{n=0}^{\infty} \frac{(-1)^n x^{2n}}{2^n \cdot n!}$$

$$\int e^{-x^2/2} \cdot dx = \int \sum_{n=0}^{\infty} \frac{(-1)^n x^{2n}}{2^n \cdot n!} \cdot dx$$

$$\int e^{-x^2/2} \cdot dx = \sum_{n=0}^{\infty} \frac{(-1)^n x^{2n+1}}{(2n+1) \cdot 2^n \cdot n!}$$

This power series is useful for computing probabilities connected with the normal distribution.

◇

# PROBLEMS

**Problem 4.65**    Find the radius of convergence for each of the following power series.

1. $\displaystyle\sum_{n=1}^{\infty} \frac{x^n}{n}$

2. $\displaystyle\sum_{n=0}^{\infty} \frac{x^n}{5^n}$

3. $\displaystyle\sum_{n=0}^{\infty} \frac{x^n}{(2n+1)!}$

4. $\displaystyle\sum_{n=0}^{\infty} \frac{x^{2n+1}}{n!}$

5. $\displaystyle\sum_{n=1}^{\infty} n x^n$

6. $\displaystyle\sum_{n=0}^{\infty} \frac{n x^n}{3n+1}$

7. $\displaystyle\sum_{n=0}^{\infty} \frac{x^n}{2n^2+4}$

8. $\displaystyle\sum_{n=0}^{\infty} \frac{x^n}{n^3+1}$

**Problem 4.66**    Demonstrate that the radius of convergence of

$$\sum_{n=0}^{\infty} a_n x^n \text{ and } \sum_{n=0}^{\infty} c a_n x^n$$

are the same for any constant $c$.

**Problem 4.67**    Compute the radius of convergence of

$$\sum_{n=1}^{\infty} \frac{x^n}{c^n}$$

where $c > 0$ is a constant.

**Problem 4.68**    If

$$f(x) = \frac{1}{1-x} + \frac{1}{1-2x} + \frac{1}{1-3x}$$

find and simplify a power series for $f(x)$.

**Problem 4.69**    If

$$g(x) = \frac{x}{2-x} + \frac{x}{3-x}$$

find and simplify a power series for $g(x)$.

**Problem 4.70**    If

$$h(x) = \frac{x^2}{2-x} + \frac{x^2}{1-2x}$$

find and simplify a power series for $h(x)$.

**Problem 4.71**    Find the interval of convergence for each of the following power series.

1. $\displaystyle\sum_{n=0}^{\infty} \frac{x^n}{4^n}$

3. $\displaystyle\sum_{n=0}^{\infty} \frac{nx^n}{2n^3 + 7}$

5. $\displaystyle\sum_{n=0}^{\infty} \frac{x^n}{n^n}$

2. $\displaystyle\sum_{n=1}^{\infty} \frac{x^n}{(2n+1)!}$

4. $\displaystyle\sum_{n=2}^{\infty} \frac{n^2 x^n}{n^4 - 1}$

6. $\displaystyle\sum_{n=0}^{\infty} \frac{(-2x)^n}{n}$

**Problem 4.72**    Using calculus, find a power series for:

$$g(x) = \ln\left(\frac{1+x}{1-x}\right)$$

**Problem 4.73**    Using only algebra, and known series, find a power series for:

$$h(x) = \frac{3x^2}{1+x^3}$$

**Problem 4.74**    Find a power series for:

$$q(x) = \ln\left(1 + x^3\right)$$

**Problem 4.75**    Find a power series for:

$$q(x) = \frac{1}{x^2 - 9}$$

**Problem 4.76**   Find a power series for:

$$q(x) = \frac{1}{x^2 - 3x + 2}$$

**Problem 4.77**   Find a power series for:

$$q(x) = x^2 \cdot \ln\left(1 + x^3\right)$$

**Problem 4.78**   Find a power series for:

$$q(x) = \frac{x^2 + 9}{x^2 - 9}$$

## 4.4   TAYLOR SERIES

In the last section, we managed to create power series for several of the standard transcendental functions. Notably absent were $\sin(x)$, $\cos(x)$, and $e^x$. The key to these is **Taylor series**. We need a little added notation to build Taylor series. We will denote the $n$th derivative of $f(x)$ by $f^{(n)}(x)$. Notice that this means that $f^{(0)}(x) = f(x)$.

### Knowledge Box 4.23

**Taylor series**

*If $f(x)$ is a function that can be differentiated any number of times,*

$$f(x) = \sum_{n=0}^{\infty} \frac{f^{(n)}(c)(x - c)^n}{n!}.$$

*This formula is called the* **Taylor series expansion of $f(x)$ at c**. *The constant $c$ is called the* **center** *of the expansion.*

**Example 4.79**   Use Taylor's formula to find a power series centered at $c = 0$ for $f(x) = e^x$ and find its radius of convergence.

**Solution:**

The function $f(x) = e^x$ is a very good choice for a first demonstration of the Taylor expansion. This is because *every* derivative of of $e^x$ is $e^x$. In other words,

$$f^{(n)}(x) = e^x \text{ and so } f^{(n)}(0) = 1$$

Applying the formula we get:

$$e^x = \sum_{n=0}^{\infty} \frac{f^{(n)}(0)(x-0)^n}{n!} = \sum_{n=0}^{\infty} \frac{1 \cdot x^n}{n!} = \sum_{n=0}^{\infty} \frac{x^n}{n!}$$

The radius of convergence of this series was computed in Example $4.56$ – it is $r = \infty$. This means that the expansion of $e^x$ converges everywhere.

◇

Here is an interesting calculation. Recall Euler's identity from *Fast Start Differential Calculus*: $e^{ix} = i\sin(x) + \cos(x)$. Let's look at the Taylor series for $e^{ix}$:

$$e^{ix} = \sum_{n=0}^{\infty} \frac{(ix)^n}{n!}$$

$$= \sum_{n=0}^{\infty} \frac{i^n x^n}{n!}$$

$$= 1 + ix - \frac{1}{2}x^2 - i\frac{1}{3!}x^3 + \frac{1}{4!}x^4 + i\frac{1}{5!}x^5 - \frac{1}{6!}x^6 - i\frac{1}{7!}x^7 + \cdots$$

$$= \sum_{n=0}^{\infty} \frac{(-1)^n x^{2n}}{(2n)!} + i\sum_{n=0}^{\infty} \frac{(-1)^n x^{2n+1}}{(2n+1)!}$$

From Euler's identity, we get that the real part of the expression above is cosine, and the imaginary part is sine. This gives us power series for $\sin(x)$ and $\cos(x)$.

**Knowledge Box 4.24**

**Taylor series for $\sin(x)$ and $\cos(x)$ and $e^x$**

$$e^x = \sum_{n=0}^{\infty} \frac{x^n}{n!}$$

$$\sin(x) = \sum_{n=0}^{\infty} \frac{(-1)^n x^{2n+1}}{(2n+1)!}$$

$$\cos(x) = \sum_{n=0}^{\infty} \frac{(-1)^n x^{2n}}{(2n)!}$$

*with all three expansions having an interval of convergence of $(-\infty, \infty)$.*

The calculations above show that we can use algebraic manipulation to create power series based on the power series we get from Taylor expansions. Let's do a couple more examples.

**Example 4.80**    Find a power series for $h(x) = e^{2x}$.

**Solution:**

$$e^x = \sum_{n=0}^{\infty} \frac{x^n}{n!}$$

$$e^{2x} = \sum_{n=0}^{\infty} \frac{(2x)^n}{n!}$$

$$= \sum_{n=0}^{\infty} \frac{2^n x^n}{n!}$$

$$= \sum_{n=0}^{\infty} \frac{2^n}{n!} x^n$$

$\Diamond$

**Example 4.81**    Find the Taylor expansion for $s(x) = \cos(x)$ using $c = \dfrac{\pi}{2}$.

**Solution:**

The Taylor formula needs the nth derivatives at $\dfrac{\pi}{2}$. Let's start by computing these numbers.

| $n$ | $s^{(n)}(x)$ | $s^{(n)}\left(\frac{\pi}{2}\right)$ |
|---|---|---|
| 0 | $\cos(x)$ | 0 |
| 1 | $-\sin(x)$ | -1 |
| 2 | $-\cos(x)$ | 0 |
| 3 | $\sin(x)$ | 1 |
| 4 | $\cos(x)$ | 0 |
| 5 | $-\sin(x)$ | -1 |

Which is enough to notice the values repeat every four steps.

Plug these values into the Taylor expansion formula with $c = \pi/2$ and we get:

$$\cos(x) = -\frac{(x - \pi/2)}{1!} + \frac{(x - \pi/2)^3}{3!} - \frac{(x - \pi/2)^5}{5!} + \frac{(x - \pi/2)^7}{7!}$$

$$\cos(x) = \sum_{n=0}^{\infty} \frac{(-1)^{n+1}(x - \pi/2)^{2n+1}}{(2n + 1)!}$$

Notice that the expansion of $\cos(x)$ with a center of $c = \dfrac{\pi}{2}$ has the same form as the negative of the Taylor expansion of $-\sin(x)$ at $c = 0$. This is a fairly extreme way of proving the identity

$$\cos(x) = -\sin(x - \pi/2)$$

### 4.4.1    TAYLOR POLYNOMIALS

If we take the first $n$ terms of a Taylor series for $f(x)$ the result is called the *Taylor polynomial of degree n* for $f(x)$. Taylor polynomials are approximations to the function they are derived from and, as we already know, polynomials are among the very nicest functions to work with.

**Example 4.82**   Find a fifth-degree Taylor polynomial for $f(x) = \sin(x)$ centered at $c = 0$.

**Solution:**

Since the Taylor series for $\sin(x)$ centered at $c = 0$ is

$$\sum_{n=0}^{\infty} \frac{(-1)^n\, x^{2n+1}}{(2n+1)!}$$

we need only extract the terms of degree five or less from the infinite series. This gives us a solution of

$$p(x) = x - \frac{x^3}{6} + \frac{x^5}{120}$$

$$\diamond$$

A natural question is "how good is this polynomial as an approximation to $\sin(x)$? Let's graph both functions on the same set of axes.

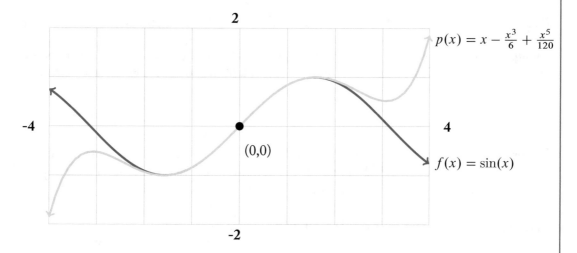

In the range $-2 \le x \le 2$ the polynomial and the sine function agree really well – after that they diverge from one another and the polynomial shoots off to positive and negative infinity. Not too surprisingly, the Taylor polynomial does a good job of approximating the function it was drawn from near the center $c$ for expansion. How do we figure out where the polynomial is a good enough approximation?

If $T(x)$ is the Taylor series for a function $f(x)$ centered at $c$ and $T_n(x)$ is the $n$th degree Taylor polynomial for that function, then we set $R_n(x) = T(x) - T_n(x)$ to get the *remainder* for the polynomial. Algebraically rearranging the terms:

$$T(x) = T_n(x) + R_n(x)$$

With this definition we can give Taylor's inequality for the remainder of a power series.

### Knowledge Box 4.25

### Taylor's Inequality

*Suppose we are looking at a Taylor polynomial for the function $f(x)$ in the interval $|x - a| \leq d$ and that $|f^{(n+1)}(x)| \leq M$ everywhere in this interval. Then for values of $x$ in the interval we have that*

$$|R_n(x)| \leq \frac{M}{(n+1)!}|x - a|^{n+1}.$$

It is not immediately obvious how to use this result, so let's do a couple examples.

**Example 4.83**    Find a bound on the error of approximation of $T_5(x)$ for $f(x) = \sin(x)$ on the interval $-2 \leq x \leq 2$. Note that this is the interval that looked good in the graph associated with Example 4.82.

**Solution:**

We proceed by applying Taylor's inequality. The derivative of $\sin(x)$ are all sine and cosine functions. This means that $f^{(n+1)}(x)$ are always at most 1. So, we may set $M = 1$ for Taylor's inequality. The value of $a$ is zero so $|x - a| \leq 2$ on the interval we are using.

Plugging these values into the formula for Taylor's inequality we get:

$$|R_5(x)| \leq \frac{1}{6!} \cdot 2^6 = \frac{64}{720} \cong 0.089$$

The error of $T_5(x)$ on $-2 \leq x \leq 2$ is *at most* 0.089. Not bad.

Notice that Taylor's inequality has $(n + 1)!$ in the denominator. That means, if we use

$$T_7(x) = x - \frac{x^3}{6} + \frac{x^5}{120} - \frac{x^7}{5040}$$

for $f(x) = \sin(x)$ on the interval $-2 \leq x \leq 2$, then the estimate of maximum error drops to

$$|R_7(x)| \leq \frac{1}{8!} \cdot 2^8 = \frac{256}{40320} \cong 0.0063$$

The factorial in the denominator lets the error drop really quickly.

$$\Diamond$$

The hard part of using Taylor's inequality is finding the constant $M$. Standard practice is to simply find the largest value of $f^{n+1}(x)$ in the interval and live with it. The fact that both sine and cosine are bounded in absolute value by 1 make them especially easy functions to work with. Let's do an example with an exponential function.

**Example 4.84**   Suppose we are approximating $f(x) = e^x$ in the interval $-3 \leq x \leq 3$ with $T_n(x)$. What value of $n$ makes the Taylor's inequality estimate of error at most 0.1?

**Solution:**

If $f(x) = e^x$ then $f^{(n+1)} = e^x$, which is very convenient. We again have that $a = 0$. So, it is pretty easy to see that

$$|f^{(n+1)}(x)| \leq e^3$$

everywhere on the interval. So, we set $M = e^3$. The largest value of $|x - a|$ on the interval is 3, so we need to find the smallest value of $n$ that makes

$$\frac{e^3}{(n+1)!} 3^{n+1} < 0.1$$

The simplest way to do this is to tabulate.

| $n$ | $|R_n(x)| \leq |\frac{e^3}{(n+1)!} \cdot 3^{n+1}|$ | $n$ | $|R_n(x)| \leq |\frac{e^3}{(n+1)!} \cdot 3^{n+1}|$ |
|---|---|---|---|
| 1 | 90.3849161543 | 6 | 8.7156883435 |
| 2 | 90.3849161543 | 7 | 3.2683831288 |
| 3 | 67.7886871158 | 8 | 1.0894610429 |
| 4 | 40.6732122695 | 9 | 0.3268383129 |
| 5 | 20.3366061347 | 10 | 0.0891377217 |

So the first value of $n$ with an acceptable error is $n = 10$, and $T_{10}(x)$ is good enough to approximate $f(x) = e^x$ in the range $-3 \le x \le 3$.

◊

The error estimates given by Taylor's inequality are not the best possible – they are actually fairly conservative. They usually over-estimate the error. If you take a course in numerical analysis later in your career you may study methods for building better error estimates. There is also a lot of room to be clever with how you use Taylor polynomials. The sine and cosine function are periodic, and so if you know their values even on a very small interval, like $[0, \pi/2]$, you can use those values to deduce any other value.

## PROBLEMS

**Problem 4.85**    Using the Taylor series formula, verify the formula for $y = \sin(x)$.

**Problem 4.86**    Using the Taylor series formula, verify the formula for $y = \cos(x)$.

**Problem 4.87**    Find the Taylor expansion of $f(x) = \sin(x)$ using a center of $c = \pi$. Use the formula for Taylor expansion.

**Problem 4.88**    Find the Taylor expansion of $f(x) = \sin(x)$ using a center of $c = \pi/2$. Use the formula for the Taylor expansion.

**Problem 4.89**    Using any method, find power series for the following functions.

1. $f(x) = e^{-x}$

2. $g(x) = xe^{2x}$

3. $h(x) = \cos 2x$

4. $r(x) = \sin(x^2)$

5. $s(x) = \ln(x^4)$

6. $q(x) = (e^x + e^{-x})/2$

7. $a(x) = \tan^{-1}(3x)$

8. $b(x) = \dfrac{3x^2}{1 - x^4}$

**Problem 4.90**   Find the Taylor expansion of $f(x) = \cos(x)$ using a center of $c = \pi/4$. Use the formula for the Taylor expansion. Warning: this is a little messy.

**Problem 4.91**   Using the Taylor series for $\sin(x)$, $\cos(x)$ and $e^x$, prove Euler's identity:

$$e^{i\theta} = i\sin(x) + \cos(x)$$

**Problem 4.92**   For each of the series you found in Problem 4.89, find the radius and interval of convergence.

**Problem 4.93**   Prove that the Taylor series for a polynomial function $p(x)$ is just the polynomial itself.

**Problem 4.94**   Find a power series expansion for

$$f(x) = \frac{1}{x^2 - 3x + 2}$$

**Problem 4.95**   Find a power series expansion for

$$f(x) = \frac{1}{4 - 4x + x^2}$$

**Problem 4.96**   Find a power series expansion for

$$f(x) = \frac{1}{x^3 - 6x^2 + 11x + 6}$$

**Problem 4.97**   Find a power series expansion for

$$f(x) = \frac{1}{x^3 + x}$$

**Problem 4.98**   If

$$f(x) = p(x)e^x$$

where $p(x)$ is a polynomial, demonstrate that $f(x)$ has a power series expansion with radius of convergence $r = \infty$.

**Problem 4.99**   Find the Taylor polynomial of degree $n$ for the given function with the given center $c$.

1. $f(x) = \cos(x)$ for $n = 6$ at $c = 0$,

2. $g(x) = \sin(2x)$ for $n = 7$ at $c = 0$,

3. $h(x) = e^x$ for $n = 5$ at $c = 0$,

4. $r(x) = \log(x)$ for $n = 3$ at $c = 1$,

5. $s(x) = \tan^{-1}(x)$ for $n = 8$ at $c = 0$,

6. $q(x) = x^2 + 3x + 5$ for $n = 2$ at $c = 1$.

**Problem 4.100**   Suppose we have $T_5(x)$ for $f(x) = e^x$ at $c = 0$. Compute a bound on the size of $R_5(x)$ with Taylor's inequality.

**Problem 4.101**   Find the smallest $n$ for which $T_n(x)$ on $f(x) = \cos(x)$ has $|R_n(x)| < 0.01$ on $-3 \le x \le 3$ with $c = 0$.

# APPENDIX A

# Useful Formulas

## A.1 POWERS, LOGS, AND EXPONENTIALS

### RULES FOR POWERS

- $a^{-n} = \dfrac{1}{a^n}$

- $\dfrac{a^n}{a^m} = a^{n-m}$

- $a^n \times a^m = a^{n+m}$

- $(a^n)^m = a^{n \times m}$

### LOG AND EXPONENTIAL ALGEBRA

- $b^{\log_b(c)} = c$

- $\log_b(x^y) = y \cdot \log_b(x)$

- $\log_b(b^a) = a$

- $\log_b(xy) = \log_b(x) + \log_b(y)$

- $\log_c(x) = \dfrac{\log_b(x)}{\log_b(c)}$

- $\log_b\left(\dfrac{x}{y}\right) = \log_b(x) - \log_b(y)$

- If $\log_b(c) = a$, then $c = b^a$

## A.2 TRIGONOMETRIC IDENTITIES

### TRIG FUNCTION DEFINITIONS FROM SINE AND COSINE

- $\tan(\theta) = \dfrac{\sin(\theta)}{\cos(\theta)}$

- $\tan(\theta) = \dfrac{1}{\cot(\theta)}$

- $\csc(\theta) = \dfrac{1}{\sin(\theta)}$

- $\cot(\theta) = \dfrac{\cos(\theta)}{\sin(\theta)}$

- $\sec(\theta) = \dfrac{1}{\cos(\theta)}$

### PERIODICITY IDENTITIES

- $\sin(x + 2\pi) = \sin(x)$

- $\sec(x) = \csc\left(x + \frac{\pi}{2}\right)$

- $\sin(x + \pi) = -\sin(x)$

- $\cos(x + 2\pi) = \cos(x)$

- $\cos(-x) = \cos(x)$

- $\cos(x + \pi) = -\cos(x)$

- $\sin(x) = \cos\left(x - \frac{\pi}{2}\right)$

- $\sin(-x) = -\sin(x)$

- $\tan(x + \pi) = \tan(x)$

- $\tan(x) = -\cot\left(x - \frac{\pi}{2}\right)$

- $\tan(x) = -\tan(x)$

## THE PYTHAGOREAN IDENTITIES

- $\sin^2(\theta) + \cos^2(\theta) = 1$     • $\tan^2(\theta) + 1 = \sec^2(\theta)$     • $1 + \cot^2(\theta) = \csc^2(\theta)$

## THE LAW OF SINES, THE LAW OF COSINES

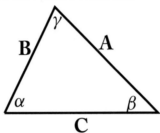

### The Law of Sines

$$\frac{A}{\sin(\alpha)} = \frac{B}{\sin(\beta)} = \frac{C}{\sin(\gamma)}$$

### The Law of Cosines

$$C^2 = A^2 + B^2 + 2AB \cdot \cos(\gamma)$$

The laws refer to the diagram.

## SUM, DIFFERENCE, AND DOUBLE ANGLE

- $\sin(\alpha + \beta) = \sin(\alpha)\cos(\beta) + \sin(\beta)\cos(\alpha)$     • $\cos(2\theta) = \cos^2(\theta) - \sin^2(\theta)$

- $\cos(\alpha + \beta) = \cos(\alpha)\cos(\beta) - \sin(\alpha)\sin(\beta)$

- $\sin(\alpha - \beta) = \sin(\alpha)\cos(\beta) - \sin(\beta)\cos(\alpha)$     • $\cos^2(\theta/2) = \dfrac{1 + \cos(\theta)}{2}$

- $\cos(\alpha - \beta) = \sin(\alpha)\sin(\beta) + \cos(\alpha)\cos(\beta)$

- $\sin(2\theta) = 2\sin(\theta)\cos(\theta)$     • $\sin^2(\theta/2) = \dfrac{1 - \cos(\theta)}{2}$

## A.3    SPEED OF FUNCTION GROWTH

- Logarithms grow faster than constants.

- Positive powers of $x$ grow faster than logarithms.

- Larger positive powers of $x$ grow faster than smaller positive powers of $x$.

- Exponentials (with positive exponents) grow faster than positive powers of $x$.

- Exponentials with larger exponents grow faster than those with smaller exponents.

# A.4   DERIVATIVE RULES

- If $f(x) = x^n$ then
$$f'(x) = nx^{n-1}$$

- $(f(x) + g(x))' = f'(x) + g'(x)$

- $(a \cdot f(x))' = a \cdot f'(x)$

- If $f(x) = \ln(x)$, then $f'(x) = \dfrac{1}{x}$

- If $f(x) = \log_b(x)$, then $f'(x) = \dfrac{1}{x \ln(b)}$

- If $f(x) = e^x$, then $f'(x) = e^x$

- If $f(x) = a^x$, then $f'(x) = \ln(a) \cdot a^x$

- $(\sin(x))' = \cos(x)$

- $(\cos(x))' = -\sin(x)$

- $(\tan(x))' = \sec^2(x)$

- $(\cot(x))' = -\csc^2(x)$

- $(\sec(x))' = \sec(x)\tan(x)$

- $(\csc(x))' = -\csc(x)\cot(x)$

- $\left(\sin^{-1}(x)\right)' = \dfrac{1}{\sqrt{1-x^2}}$

- $\left(\cos^{-1}(x)\right)' = \dfrac{-1}{\sqrt{1-x^2}}$

- $\left(\tan^{-1}(x)\right)' = \dfrac{1}{1+x^2}$

- $\left(\cot^{-1}(x)\right)' = \dfrac{-1}{1+x^2}$

- $\left(\sec^{-1}(x)\right)' = \dfrac{1}{|x|\sqrt{x^2-1}}$

- $\left(\csc^{-1}(x)\right)' = \dfrac{-1}{|x|\sqrt{x^2-1}}$

**The product rule**
$$(f(x) \cdot g(x))' = f(x)g'(x) + f'(x)g(x)$$

**The quotient rule**
$$\left(\frac{f(x)}{g(x)}\right)' = \frac{g(x)f'(x) - f(x)g'(x)}{g^2(x)}$$

**The reciprocal rule**
$$\left(\frac{1}{f(x)}\right)' = \frac{-f'(x)}{f^2(x)}$$

**The chain rule**
$$(f(g(x)))' = f'(g(x)) \cdot g'(x)$$

# A.5   SUMS AND FACTORIZATION RULES

## FACTORIZATIONS OF POLYNOMIALS

- $x^2 - a^2 = (x - a)(x + a)$

- $x^3 - a^3 = (x - a)(x^2 + ax + a^2)$

- $x^3 + a^3 = (x + a)(x^2 - ax + a^2)$

- $x^n - a^n = (x - a)(x^{n-1} + ax^{n-2} + \cdots a^{n-2}x + a^{n-1})$

## ALGEBRA OF SUMMATION

- $\displaystyle\sum_{i=a}^{b} f(i) + g(i) = \sum_{i=a}^{b} f(i) + \sum_{i=a}^{b} g(i)$ 
- $\displaystyle\sum_{i=a}^{b} c \cdot f(i) = c \cdot \sum_{i=a}^{b} f(i)$

## CLOSED SUMMATION FORMULAS

- $\displaystyle\sum_{i=1}^{n} 1 = n$
- $\displaystyle\sum_{i=1}^{n} i^2 = \frac{n(n+1)(2n+1)}{6}$

- $\displaystyle\sum_{i=1}^{n} i = \frac{n(n+1)}{2}$
- $\displaystyle\sum_{i=1}^{n} i^3 = \frac{n^2(n+1)^2}{4}$

## A.5.1   GEOMETRIC SERIES

- $\displaystyle\sum_{k=0}^{n} a^k = \frac{a^{n+1} - 1}{a - 1}$
- $\displaystyle\sum_{n=0}^{\infty} qa^n = \frac{q}{1 - a}$ if $|a| < 1$

- $\displaystyle\sum_{n=0}^{\infty} a^n = \frac{1}{1 - a}$ if $|a| < 1$

# A.6 VECTOR ARITHMETIC

## VECTOR ARITHMETIC AND ALGEBRA

- $c \cdot \vec{v} = (cv_1, cv_2, \ldots, cv_n)$
- $\vec{v} + \vec{w} = (v_1 + w_1, v_2 + w_2, \ldots, v_n + w_n)$
- $\vec{v} - \vec{w} = (v_1 - w_1, v_2 - w_2, \ldots, v_n - w_n)$
- $\vec{v} \cdot \vec{w} = v_1 w_1 + v_2 w_2 + \ldots + v_n w_n$

- $c \cdot (\vec{v} + \vec{w}) = c \cdot \vec{v} + c \cdot \vec{w}$
- $c \cdot (d \cdot \vec{v}) = (cd) \cdot \vec{v}$
- $\vec{v} + \vec{w} = \vec{w} + \vec{v}$
- $\vec{u} \cdot (\vec{v} + \vec{w}) = \vec{u} \cdot \vec{v} + \vec{u} \cdot \vec{w}$

## CROSS PRODUCT OF VECTORS

- $\vec{v} \times \vec{w} = (v_2 w_3 - v_3 v_2, \ v_3 w_1 - v_1 w_3, \ v_1 w_2 - v_2 w_1)$

**Formula for the angle between vectors**

$$\cos(\theta) = \frac{\vec{v} \cdot \vec{w}}{|v||w|}$$

# A.7 POLAR AND RECTANGULAR CONVERSION

- $x = r \cdot \cos(\theta)$
- $y = r \cdot \sin(\theta)$

- $r = \sqrt{x^2 + y^2}$
- $\theta = \tan^{-1}(y/x)$

# A.8 INTEGRAL RULES

## BASIC INTEGRATION RULES

- $\int x^n \cdot dx = \frac{1}{n+1} x^{n+1} + C$
- $\int (f(x) + g(x)) \cdot dx = \int f(x) \cdot dx + \int g(x) \cdot dx$

- $\int a \cdot f(x) \cdot dx = a \cdot \int f(x) \cdot dx$

## LOG AND EXPONENT

- $\int \frac{1}{x} \cdot dx = \ln(x) + C$
- $\int e^x \cdot dx = e^x + C$

# TRIG AND INVERSE TRIG

- $\displaystyle\int \sin(x)\cdot dx = -\cos(x) + C$

- $\displaystyle\int \cos(x)\cdot dx = \sin(x) + C$

- $\displaystyle\int \sec^2(x)\cdot dx = \tan(x) + C$

- $\displaystyle\int \csc^2(x)\cdot dx = -\cot(x) + C$

- $\displaystyle\int \sec(x)\tan(x)\cdot dx = \sec(x) + C$

- $\displaystyle\int \csc(x)\cot(x)\cdot dx = -\csc(x) + C$

- $\displaystyle\int \frac{1}{\sqrt{1-x^2}}\cdot dx = \sin^{-1}(x) + C$

- $\displaystyle\int \frac{1}{1+x^2}\cdot dx = \tan^{-1}(x) + C$

- $\displaystyle\int \frac{1}{x\sqrt{x^2-1}}\cdot dx = \sec^{-1}(|x|) + C$

- $\displaystyle\int \tan(x)\cdot dx = \ln|\sec(x)| + C$

- $\displaystyle\int \sec(x)\cdot dx = \ln|\sec(x) + \tan(x)| + C$

# INTEGRATION BY PARTS

$$\int U\cdot dV = UV - \int V\cdot dU$$

# EXPONENTIAL/POLYNOMIAL SHORTCUT

$$\int p(x)e^x\cdot dx = \left(p(x) - p'(x) + p''(x) - p'''(x) + \cdots\right)e^x + C$$

# VOLUME, SURFACE, ARC LENGTH

**Volume of Rotation, Disks**

$$V = \pi\int_a^b f(x)^2\cdot dx$$

**Volume of Rotation, Cylindrical Shells**

$$V = 2\pi\int_{x=a}^{x=b} x\cdot f(x)\cdot dx$$

**Volume of Rotation with Washers**

$$V = \pi\int_a^b \left(f_1(x)^2 - f_2(x)^2\right)\cdot dx$$

**Differential of Arc Length**

$$ds = \sqrt{(y')^2}\cdot dx = \sqrt{(f'(x))^2}\cdot dx$$

**Arc Length**

$$S = \int ds$$

**Surface Area of Rotation**

$$A = 2\pi\int_a^b f(x)\cdot ds$$

# A.9   SERIES CONVERGENCE TESTS

## DIVERGENCE TEST

If $\lim\limits_{n\to\infty} x_n \neq 0$ then $\sum\limits_{n=0}^{\infty} x_n$ does not have a finite value.

## INTEGRAL TEST

Suppose that $f(x)$ is a positive, decreasing function on $[0, \infty)$ and that $x_n = f(n)$. Then $\sum\limits_{n=1}^{\infty} x_n$ and $\int_{a}^{\infty} f(x)dx$ both converge or both diverge for any finite $a \geq 1$.

## P-SERIES TEST

The $p$-series $\sum\limits_{n=1}^{\infty} \dfrac{1}{n^p}$ converges if $p > 1$ and diverges if $p \leq 1$.

## COMPARISON TESTS

- If $\sum\limits_{n=0}^{\infty}$ converges and $0 \leq y_n \leq x_n$ for all $n$, then $\sum\limits_{n=0}^{\infty} y_n$ also converges.

- If $\sum\limits_{n=0}^{\infty} x_n$ diverges and $y_n \geq x_n \geq 0$ for all $n$, then $\sum\limits_{n=0}^{\infty} y_n$ also diverges.

## LIMIT COMPARISON TEST

Suppose that $\sum\limits_{n=0}^{\infty} x_n$ and $\sum\limits_{n=0}^{\infty} y_n$ are series, and $\lim\limits_{n\to\infty} \dfrac{x_n}{y_n} = C$ where $0 < |C| < \infty$ is a constant. Then, either both series converge or both series diverge.

## ABSOLUTE CONVERGENCE TEST

If $\sum\limits_{n=0}^{\infty} |x_n|$ converges, then so does $\sum\limits_{n=0}^{\infty} x_n$.

## ALTERNATING SERIES TEST

Suppose that $x_n$ is a series such that $x_n > x_{n+1} \geq 0$, and suppose that $\lim\limits_{n\to\infty} x_n = 0$. Then the series $\sum\limits_{n=0}^{\infty} (-1)^n \cdot x_n$ converges.

## RATIO TEST

Suppose that $\sum_{n=0}^{\infty} x_n$ is a series and $r = \lim_{n \to \infty} \left| \frac{x_{n+1}}{x_n} \right|$. Then

- if $r < 1$, then the series converges
- if $r > 1$, then the series diverges
- if $r = 1$, then the test is inconclusive

## ROOT TEST

Suppose that $\sum_{n=0}^{\infty} x_n$ is a series and $s = \lim_{n \to \infty} \sqrt[n]{|x_n|}$. Then

- if $s < 1$, then the series converges
- if $s > 1$, then the series diverges
- if $s = 1$, then the test is inconclusive

## A.10  TAYLOR SERIES

If $f(x)$ is a function that can be differentiated any number of times,

$$f(x) = \sum_{n=0}^{\infty} \frac{f^{(n)}(c)(x-c)^n}{n!}$$

### SPECIAL TAYLOR SERIES

- $e^x = \sum_{n=0}^{\infty} \frac{x^n}{n!}$        - $\sin(x) = \sum_{n=0}^{\infty} \frac{(-1)^n x^{2n+1}}{(2n+1)!}$        - $\cos(x) = \sum_{n=0}^{\infty} \frac{(-1)^n x^{2n}}{(2n)!}$

### TAYLOR'S INEQUALITY

Suppose we are looking at a Taylor polynomial for the function $f(x)$ in the interval $|x - a| \leq d$ and that $|f^{(n+1)}(x)| \leq M$ everywhere in this interval. Then for values of $x$ in the interval we have that

$$|R_n(x)| \leq \frac{M}{(n+1)!}|x - a|^{n+1}$$

# Author's Biography

## DANIEL ASHLOCK

**Daniel Ashlock** is a Professor of Mathematics at the University of Guelph. He has a Ph.D. in Mathematics from the California Institute of Technology, 1990, and holds degrees in Computer Science and Mathematics from the University of Kansas, 1984. Dr. Ashlock has taught mathematics at levels from 7th grade through graduate school for four decades, starting at the age of 17. Over this time Dr. Ashlock has developed a number of ideas about how to help students overcome both fear and deficient preparation. This text, covering the mathematics portion of an integrated mathematics and physics course, has proven to be one of the more effective methods of helping students learn mathematics with physics serving as an ongoing anchor and example.

# Index